PHYSICS 3
WORKBOOK
for NCEA Level 3

Rob Boasman

Physics 3 Workbook for NCEA Level 3
1st Edition
Rob Boasman

Production controller: Siew Han Ong
Typeset by Cheryl Rowe, Macarn Design

Any URLs contained in this publication were checked for currency during the production process. Note, however, that the publisher cannot vouch for the ongoing currency of URLs.

Thanks

Having worked as a bank clerk, jazz musician, mine shaft electrical engineer, builders mate, power salesman, and getaway car driver, I can honestly say that having the opportunity to see the 'ahh!' moment in a student's face makes teaching the most uniquely rewarding profession. As a teacher for over twenty years my passion for physics and astronomy are stronger than ever thanks to the thousands of students who have challenged me to look deeper into this fascinating subject. I hope that this book will not only meet your learning needs but provide you with a glimpse behind the watch face of the universe.

I am indebted to the following people who have helped me prepare this book: Jenny Thomas, Graham McEwan and Eva Chan at Cengage and Cheryl Smith at Macarn Design who have demonstrated tremendous patience; Ashwin Reddy and Carol TeBay who spent many hours proof reading and making me less ignorant of my ignorance; the teachers and students at Diocesan School for Girls who have tirelessly pointed out typos and eaten all my Skittles; Heather McRae the Principal at Diocesan School for Girls who has been a tremendous support; Lorraine Rose for keeping my family and I sane during the writing of this book; Richard Feynman, Albert Einstein and Terry Pratchett for revealing the low fat side of life; all the readers of my L2 books who have encouraged me to keep going; and finally my wife Sarah, and sons Joel, Josh and Sam for forbearance, support, encouragement and keeping the quantum butterfly under control!

Rob Boasman
Diocesan School for Girls
Auckland
2015

For product information and technology assistance,
in Australia call **1300 790 853**;
in New Zealand call **0800 449 725**

For permission to use material from this text or product, please email **aust.permissions@cengage.com**

National Library of New Zealand Cataloguing-in-Publication Data
A catalogue record for this book is available from the National Library of New Zealand.

ISBN 978 017 0 368179

Cengage Learning Australia
Level 7, 80 Dorcas Street
South Melbourne, Victoria Australia 3205

Cengage Learning New Zealand
Unit 4B Rosedale Office Park
331 Rosedale Road, Albany, North Shore 0632, NZ

For learning solutions, visit **cengage.co.nz**

Printed in China by 1010 Printing International Limited
8 9 10 25 24

Appendices and Answers
Appendices and a full set of comprehensive answers at:
www.nelsonsecondary.co.nz/physics3

CONTENTS

1 Understanding the physical world

Communicating in science

The skills, attitudes, values and knowledge established in Physics at Levels 1 and 2 enable students to start describing the world around them. Level 3 Physics explores the physical world in greater depth, introducing students to new vocabulary, symbols, models, rules and conventions. Students will need to use these when evaluating accounts of the natural world and connecting their ideas to historical scientific events and the latest developments. With a firm grounding in scientific language, students can develop a coherent understanding of socio-scientific issues that concern them and form a reasoned response at both a personal and societal level.

Students must be confident in the use of physical quantities, units, Greek symbols and number forms.

Physical quantities and SI units

The study of physics involves the measurement and analysis of **physical quantities**. For consistency of measurement, all quantities are measured in SI units (an abbreviation of the French *Système International d'Unités*). The seven fundamental quantities and their units are shown in the table.

Quantity		Unit	
Name	Symbol	Name	Symbol
Length	L, d, r	metre	m
Mass	m	kilogram	kg
Time	t	second	s
Temperature	T	kelvin	K
Electric current	I	ampere	A
Luminosity	L	candela	cd
Amount of a substance	n	mole	mol

Derived units

All other physical quantities are derived from the fundamental quantities listed above. The table below shows some of the Level 3 derived quantities and units that are used in this book.

Derived quantity		Unit		
Name	Symbol	Name	Symbol	Fundamental units
Charge	Q	coulomb	C	A s
Energy	E	joule	J	$kg\ m^2\ s^{-2}$
Potential difference (pd)	V	volt	V	$kg\ m^2\ s^{-3}\ A^{-1}$
Resistance	R	ohm	Ω	$kg\ m^2\ s^{-3}\ A^{-2}$
Inductance	L	henry	H	$kg\ m^2\ s^{-2}\ A^{-2}$
Magnetic flux	Φ	weber	Wb	$kg\ m^2\ s^{-2}\ A^{-1}$
Magnetic flux density	B	tesla	T	$kg\ s^{-2}\ A^{-1}$
Capacitance	C	farad	F	$kg^{-1}\ m^{-2}\ s^4\ A^2$

 ISBN: 9780170368179

Symbols

Mathematical symbols and Greek letters are frequently used when writing physics statements or equations. The table below shows some of the symbols that are used and what they mean.

Name	Symbol	Meaning/Use
Not equal to ...	$\neq$	Two sides of an equation are not the same.
Approximately equal to ...	$\approx$	Two sides of an equation can be considered to be the same to simplify a solution, e.g. for small angles, $\sin \theta \approx \theta$.
Exactly equal to ...	$\equiv$	Quantities with a precisely defined value, e.g. the mass of a carbon atom is defined as $m_{carbon} \equiv 1.2 \times 10^{-2}$ kg mol^{-1}.
Less than ...	$<$	The quantity on the left is less than the quantity on the right of an equation.
Less than or equal to ...	$\leq$	The quantity on the left is less than or equal to the quantity on the right of an equation.
Plus or minus	$\pm$	The quantity could be either positive or negative, e.g. $\sqrt{16} = \pm 4$.
Change in ...	Δ (delta)	Found by calculating the difference between the final and initial values of a changing quantity, e.g. the change in velocity $\Delta v = v_f - v_i$.
Proportional to ...	$\propto$	The change in one quantity results in the same-sized-change in another quantity, e.g. $F_{net} \propto a$, so doubling the net force will double the acceleration.
Infinity	∞	A number greater than any real number; without limit.
Angles	$\theta \quad \phi$ (theta) (phi)	Used to indicate angles.
Sum of ...	Σ (sigma)	Found by adding all relevant values together, e.g. the net force $F_{net} = \Sigma F = F_1 + F_2 + F_3 + ...$
Absolute brackets	$\lvert x \rvert$	Makes the value inside the bracket positive.

Number forms

Very large and very small numbers may be expressed in four forms: number, scientific notation, engineering notation or prefix notation.

Form	Notation	Coefficient	Exponent	Example: the speed of light, c
Number	a	a	none	299 792 458
Scientific	$a \times 10^b$	$1 \leq a < 10$	b is any whole number	2.99792458×10^8
Engineering	$a \times 10^b$	$1 \leq a < 1000$	b is any multiple of 3	299.792458×10^6

Very large and very small numbers may also be expressed using prefixes, for example centimetres (cm), where *centi* means $\times 10^{-2}$. The standard prefixes are shown below.

Prefix	Symbol	Multiplier	Example	Prefix	Symbol	Multiplier	Example
kilo	k	$\times 10^3$	km, kilometre	milli	m	$\times 10^{-3}$	mA, milliamp
mega	M	$\times 10^6$	MHz, megahertz	micro	μ	$\times 10^{-6}$	μT, microtesla
giga	G	$\times 10^9$	GW, gigawatt	nano	n	$\times 10^{-9}$	nm, nanometre
tera	T	$\times 10^{12}$	TB, terabyte	pico	p	$\times 10^{-12}$	pg, picogram

All prefixes must be converted to one of the number forms above, before they are used in equations.

ISBN: 9780170368179

Significant figures

The accuracy of any number is represented by the number of **significant figures** or 'sf'. The rules for applying significant figures are presented below.

Rule	Example	Number of significant figures	Scientific notation
All non-zero digits are significant.	43.21	4 sf	4.321×10^1
Zeros between numbers are significant.	2000.1	5 sf	2.0001×10^3
Trailing zeros after a decimal point are significant.	0.00900	3 sf	9.00×10^{-3}
Leading zeros are *not* significant.	0.0007	1 sf	7×10^{-4}
Trailing zeros after a number not containing a decimal point *may* or *may not* be significant and must be clarified by an sf statement or using scientific notation.	500	1 sf or 3 sf	5×10^2 or 5.00×10^2

When combining numbers by multiplying or dividing, the final answer should be given to the least number of significant figures in the supplied data. However, when combining numbers by adding or subtracting (for example when calculating averages), the final answer should be given to the least number of decimal places.

Angles and circles

The radian (rad) is the standard mathematical unit used to describe angles and is related to the radius of a circle. In the diagram opposite, the arc **AB** is the same length as the radius, r, of the circle, so considering the ratio of the angles and the arc lengths we have:

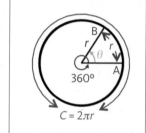

$$\frac{\theta}{360} = \frac{r}{2\pi r}$$

which gives:

$$\theta = \frac{360}{2\pi r} = \frac{180}{\pi} = 57.3° = 1 \text{ radian}$$

Conversions between degrees and radians can be done using the following formulae:

$$\theta_{rad} = \theta_{deg} \times \frac{\pi}{180} \quad \text{and} \quad \theta_{deg} = \theta_{rad} \times \frac{180}{\pi}$$

Using radians makes solving angular problems much simpler. Consider the ratio of the arc CD of length d and the angles in the circle opposite.

Using radians:	Using degrees:
$d = r\theta_{rad}$	$d = r\theta_{deg} \times \dfrac{\pi}{180}$

Written questions

The scientific method used by physicists when trying to explain the natural world starts with observations and measurements. When solving any physics problems, physics students must first identify the quantities and values that are already known, or can be measured, and then identify the quantity that is being investigated. The next step is to recognise any laws, principles or formulas that will bring the known and unknown quantities together.

 ISBN: 9780170368179

Throughout this book there are worked examples that have been presented using a strategy known as **GUESS**, which stands for:

- **G**iven — identify the given quantities and convert them into the correct units where necessary.
- **U**nknown — identify the unknown quantity.
- **E**quation — identify any equations on the formula page that link the unknown quantity to some or all of the given quantities. Where a single formula does not link all the terms, find a second formula that can link to a quantity in the first equation and any unused quantities.
- **S**ubstitute — substitute the values or changes into the formula and rearrange it where necessary.
- **S**olve — complete all the steps in the equation to solve the problem and consider whether your answer is realistic. The answer should be presented with the relevant unit and to the correct number of significant figures.

The GUESS strategy is effective for both mathematical problems and problems requiring explanation.

Physics is a bilingual subject and students must be able to use both language and mathematics skills equally competently. Appendix 1 contains a number of important mathematical formulas used in Level 3 Physics that students should be familiar with.

Exercise 1A

1   a   Using the tables on page 4, determine the unit(s) for the following combinations of quantities and hence determine the quantity that the combination can be used to calculate.

   i   current x time   _____

   ii $\dfrac{\text{energy}}{\text{charge}}$ _____

   iii $\dfrac{\text{potential difference}}{\text{resistance}}$ _____

   iv $\dfrac{\text{magnetic flux}}{\text{magnetic flux density}}$ _____

   b   Convert the following quantities into the fundamental SI unit, and express them in engineering form and scientific notation.

	Number	SI unit	Engineering notation	Scientific notation
i	2.25 min			
ii	350 nm			
iii	180.0 MΩ			
iv	0.72 T			
v	240 µF			

c   Change the following written statements into mathematical statements.

i   The speed of light, c, is **approximately equal** to 3.0×10^8 m s^{-1}.

_____

ii The total current in a circuit is equal to the **sum** of the current in each individual branch.

_____

iii The **change in** momentum is equal to the force multiplied by the **change in** time.

_____

d   Change the following mathematical statements into written statements.

i $\Delta E_p = mg\Delta h$ _____

_____

ii $f \propto \dfrac{1}{\lambda}$ _____

_____

iii $\mu \equiv 4\pi \times 10^{-7}$ T m A^{-1} _____

_____

e   Complete the table by converting the quantities to the other number forms and prefix. State how many significant figures are present in each number.

Number	Scientific notation	Engineering notation	Prefix	Significant figures
299 792 458				9 sf
	1.2566×10^{-6}			
		66.7384×10^{-12}		
			931.49 M	

f   Each of the following equations have been solved but the answers have not been recorded correctly. In each example, write the answer using an appropriate number form and to the correct number of significant figures or decimal places, and explain your decision.

i $50^2 = 2500$ _____

_____

_____

ii $317 \div 2.00 \times 10^3 = 0.1585$ _____

_____

iii $10.52 + 9.98 - 2.50 = 18$ _____

_____

    ISBN: 9780170368179

g The following equations are used in Level 3 Physics and being able to rearrange them is a valuable skill. Rearrange them in terms of the selected unknown, clearly stating what you do in each step so that someone else can follow your working.

i Rearrange $T = 2\pi\sqrt{\dfrac{l}{g}}$ to find l.

_____

_____

_____

_____

ii Rearrange $f' = f\dfrac{v_w}{v_w + v_s}$ to find v_s.

_____

_____

_____

_____

iii Rearrange $x_{CoM} = \dfrac{m_1 x_1 + m_2 x_2}{m_1 + m_2}$ to find m_2.

_____

_____

_____

_____

h Determine the following angles in radians, and state the fraction of a circle that the angle represents. The first one has been done for you.

	Angle in degrees	Angle in radians	Fraction of a circle
i	90°	$\pi/2$ rad	¼ of a circle
ii	180°		
iii	270°		
iv	360°		
v	900°		

i Complete the following table by determining the missing quantities.

	Angle in degrees	Angle in radians	Fraction of a circle
i		$\pi/4$ rad	
ii		$2\pi/3$ rad	
iii			5/6 of a circle
iv			12 circles

j   Mark each of the following angles on the circle below and then immediately plot the sine of
its angle in the graph alongside (0 and π/6 have been done for you) — you will need to put
your calculator in **radians mode** to do this. The first two have been done for you.

Angle (rad)	sin θ
0	0
π/6 rad	0.50
π/2 rad	
2π/3 rad	
π rad	
7π/6 rad	
3π/2 rad	
11π/6 rad	
2π rad	

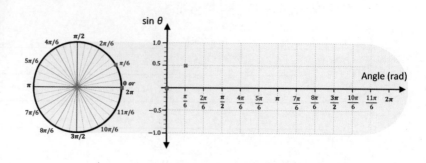

Join up the points with a smooth curve. Your graph describes how the angles in a circle
relate to a sine wave — this relationship will be extremely important later in Mechanics,
Waves and Electricity.

The following questions will require the use of **Appendix 1**.

Describing motion in two dimensions involves summing vectors and resolving vectors using
Pythagoras's theorem and trigonometry (or cosine and sine rules).

k   Calculate the missing quantities of each triangle of vectors shown below.

   i   Summing displacement vectors — right-angled triangle. Find: L, θ and φ.

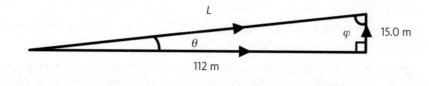

_____

_____

_____

   ii   Summing velocity vectors — non-right-angled triangle. Find: v, θ and φ.

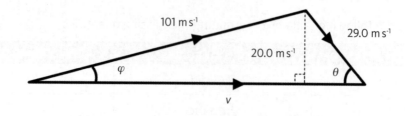

_____

_____

_____

iii Summing momentum vectors — non-right-angled triangle. Find: p and φ.

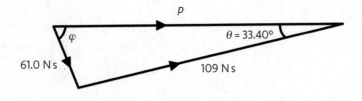

_____

_____

_____

_____

_____

_____

iv Resolving distance and speed vectors for a particle moving in a circle. Find: d_y, θ and v_y.

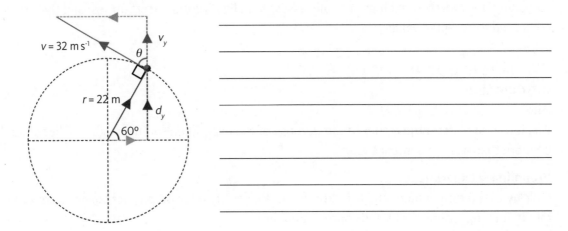

_____

_____

_____

_____

_____

_____

_____

_____

l Present the following statements in an alternative form by converting the indices, then solve, for example: $5^{-2} = \dfrac{1}{5^2} = 0.04$.

i $\dfrac{1}{3^4}$ = _____ = _____

ii $3125^{1/5}$ = _____ = _____

iii $64^{-1/6}$ = _____ = _____

iv $4^2 \times 4^5$ = _____ = _____

v $3^{-3} \times 3^3$ = _____ = _____

vi $5^5 \div 5^4$ = _____ = _____

vii $(2^6)^2$ = _____ = _____

viii $(5^6)^{1/3}$ = _____ = _____

m Present the following statements in an alternative form by converting the logarithm equations, then solve.

i $\log(6) + \log(3)$ = _____

ii $\log(20) - \log(4)$ = _____

iii $\log(\dfrac{15}{5})$ = _____

iv $\log(10^2)$ = _____

2 Experimental techniques

Through experiments and investigations, science students are able to extend their science knowledge, and develop their understanding of the relationship between investigations and scientific theories and models. Investigative work must be rigorously carried out and students must demonstrate a qualitative understanding of the uncertainties and errors that are inevitably present in all experimental work. All their work needs to be presented in such a way that their experiment could be repeated; their findings critically assessed in light of the uncertainties; and their results compared to accepted values.

Planning and gathering data

Introduction

Aim

An aim should inform the reader of the purpose of the experiment or investigation, but should not be more than a single sentence long.

Preliminary experiment

Quickly trial the equipment and the experiment to identify relevant physics theories, maximum and minimum ranges, limitations and control variables.

Hypothesis

Identify any relevant physics theories and any limitations or assumptions that apply to the investigation. Make a prediction based on these physics ideas.

Method

Independent variable

- State the independent variable (the variable being **changed** by the scientist).
- Identify the range. If the range is reasonable, the graph of the results will show the relationship between the independent and the dependent variables. (As a rough guide, about 70% or more of the measurable range should be used.)
- A minimum of five different values of the independent variable should be tested.
- Justify any limitations to the range due to physical theories, apparatus, the measurements and/ or safety.
- Describe and explain any techniques used to increase the accuracy of the measurement of the independent variable (if appropriate).
- Describe any difficulties encountered when measuring the independent variable and discuss how they were overcome.
- Determine the size of the uncertainty in the raw measurements of the independent variable.

 ISBN: 9780170368179

Dependent variable

- State the dependent variable (the variable being **measured** by the scientist).
- Describe and explain any techniques used to increase the accuracy of the measurement of the dependent variable.
- Describe any difficulties encountered when measuring the dependent variable and discuss how they were overcome.
- Determine the size of the uncertainty in the raw measurements of the dependent variable.

Control variables

- Describe any other variables that would **significantly** affect the results of the experiment if they are not kept constant, describe how they are controlled and explain how they could change the outcome if not controlled.
- Determine the size of the uncertainty in the raw measurements of any control variables.

Experimental set-up

- A fully labelled diagram of the apparatus, showing how it has been set up and how the independent and dependent measurements will be taken.

Experimental techniques and uncertainties

In experimental work, the term 'error' is often used to describe the degree of 'uncertainty' in a value due to variations in the measurement that are caused by the lack of precision in a measuring device, or the person using it. **Random** and **systematic** errors are the two main types of uncertainties that occur during an experiment.

Random errors

The table below lists some possible random errors that can occur during experimental work. Estimating the size of the uncertainty before taking any results is a good way of ensuring that the results are recorded to the correct number of significant figures. It also helps to identify anomalous readings during the investigation.

Error	Apparatus	Correction	Typical size of error
Reaction time Time taken between an event occurring and the observer reacting.	Stopwatches, clocks, mobile phones.	Repeat the measurement and calculate the average. AND/OR Multiple measurements for periodic events, e.g. a swinging pendulum. $$t_{1\ event} = \frac{t_{multiple\ events}}{number\ of\ events}$$	Estimate to be between 0.1 s and 0.2 s.
Parallax Difficulty in reading analogue meters due to the marker and the scale not being in contact.	Analogue meters, e.g. rulers, newton meters, voltmeters, ammeters.	Ensure that the eye of the observer, the marker and the scale are all in line. Some electrical meters have a mirror behind the marker to help with alignment. Repeat the measurement and calculate the average.	Estimate based on the amount the reading changes when the observer moves slightly from side to side.

(continued over page)

Error	Apparatus	Correction	Typical size of error
Counting Occurs when counting multiple events or objects.	The observer. Digital counters.	Repeat the count. OR If the event is random or spontaneous, repeat the count and calculate the average.	Eliminated by repeating the count. OR For random or spontaneous event, e.g. radioactive decay, use half the size of the range. half range error = $\dfrac{max - min}{2}$
Division Judging the position of the marker when it points between the divisions on the scale.	Analogue meters and digital meters.	Choose a scale that has smaller divisions, e.g. use the mm scale instead of the cm scale on a ruler. OR Multiple measurements for small objects to produce a value more suitable to the scale. OR Repeat the measurement and calculate the average.	Estimate as the **smallest division on the scale** being used. An analogue scale can be read with reasonable accuracy to half the smallest division. HOWEVER, as any measurement requires two readings, at the start and the end, the final error will be 2 x ½ the smallest division = 1 division.
Irregularities Density, width, length, etc. is not uniform across the object.	Objects being measured.	Repeat the measurements at several different points, e.g. measure the diameter of a ball across several different diameters and calculate the average.	Half the size of the range. half range error = $\dfrac{max - min}{2}$

Random errors, repeats and averages

Random errors vary in size and are just as likely to be positive or negative, causing each reading to be spread out around the true value. A good approximation of the 'true' value can be gained by **repeating** each measurement several times, identifying anomalous (significantly different) results and taking an **average** (mean).

$$average = \frac{trial\ 1 + trial\ 2 + trial\ 3 + ...}{number\ of\ trials}$$

The size of the uncertainty in the average can be determined in several ways:

- **Initial estimate** — reasonable approximation. The estimated error in the individual readings is taken as the error in the average. (Pros: Straightforward. Cons: Doesn't take into account the amount of uncertainty observed in the experiment.)

 ISBN: 9780170368179

- **Half range** — good approximation. Once the data has been gathered and anomalous results discounted, find the difference (range) between the maximum and minimum values of the repeat readings and halve it.

$$\text{half range error} = \frac{t_{max} - t_{min}}{2}$$

(Pros: Uses the measured data to produce a more realistic estimate of the uncertainty in the results. Cons: The uncertainties can be too large if either the maximum or minimum values are significantly different from the rest of the repeat readings.)
- **(OPTIONAL) Standard deviation** — this is the most valid method of determining the uncertainty as it takes into account the variation of all the repeat readings (not just the maximum and minimum). The standard deviation formula is quite complex, but can be easily performed with a calculator or spreadsheet program, e.g. Excel (see Appendix 2a).

The significance of the effect of random errors on an experiment can be estimated at the end of an experiment by considering the graph and the proximity (closeness) of the plotted points to the line of best fit. If all the plotted points lie close to the line of best fit, then random errors must have been small.

Systematic errors

Most measuring instruments will not be perfect and this will result in errors in the accuracy of the readings. A systematic error is one which is constant throughout the experiment, making all the readings too high or all the readings too low. Systematic errors are typically caused by the following.

Error	Apparatus	Correction
Zero error When the marker doesn't start from zero on the scale.	Analogue meters, e.g. voltmeters, ammeters, rulers, newton meters.	Some electrical meters and newton meters have an adjustment screw, which allows the position of the marker needle to be corrected. OR Measure the size of the zero error and add or subtract it (as appropriate) so that the recorded reading is correct.
Calibration error When a device has not been set up to measure correctly.	Analogue meters with incorrectly marked scales, e.g. incorrectly printed meter rulers. Low batteries on digital multimeters. Mobile phone measurement apps.	Initially confirm the readings using an alternative device(s). As it is impossible to determine which device is correct, comparing the similarity will identify any apparatus with a significant calibration error. The device considered most reliable should then be used for all readings.

The size of a systematic error can be estimated at the end of an experiment by comparing the equation of the graph line to a known physics relationship. The difference between the two equations may be due to a systematic error, and provides an opportunity to discuss possible causes.

Working with errors

Errors can be presented in three different ways: absolute (δ), relative (δ_R) and percentage ($\delta_\%$).

Absolute error (δ)

Absolute error is the estimated amount of error in the measurement based on the random and systematic errors present:

$$\text{recorded value} = x \pm \delta x$$

Absolute errors are quoted to **one significant figure**, and the decimal place of the error fixes the number of decimal places in the recorded value. For example: 15.07 ± 0.15 m is correctly written as 15.1 ± 0.2 m.

Relative error (δ_R)

Relative error compares the size of the absolute error to the measured reading.

$$\delta_R x = \frac{\delta x}{x}$$

Relative errors are used when performing error calculations (see below).

Percentage errors ($\delta_\%$)

The percentage error is determined by multiplying the relative error by 100%. Percentage errors are used to decide if an error is significant or not and it provides an idea of the reliability of the recorded value.

$$\delta_\% x = \frac{\delta x}{x} \times 100\%$$

A reliable measurement will have an error of 5% or less of the recorded value. A significant error is greater than 10%.

Error calculations

The error in any single measurement will always be the largest error of those contributing to the uncertainty in the value. When two or more measurements are combined, the errors must also be combined to determine the size of the error in the final answer. There are some simple rules to combining errors in Level 3 Physics.

Adding or subtracting	Example
When two quantities are added or subtracted to obtain a final result, the errors must **always be added** to find the largest possible combined error that could occur.	Time taken for a ball to roll between two points.
	Times: $t_i = 0.66 \pm 0.05$ s and $t_f = 1.52 \pm 0.05$ s
	Time taken, Δt:
Value: $\quad L = d + x$ or $L = d - x$	$\Delta t = t_f - t_i = 1.52 - 0.66 = 0.86$ s
Error: $\quad \delta L = \delta d + \delta x$	Error in time taken, $\delta \Delta t$:
Answer: $\quad L \pm \delta L$	$\delta \Delta t = 0.05 + 0.05 = 0.10$ s
Note: To add or subtract two or more numbers, they **must have the same unit**, i.e. be the same quantity.	Final answer:
	$\Delta t = 0.9 \pm 0.1$ s

 ISBN: 9780170368179

Average (mean)	Example
Calculating the mean value from a set of trials won't reveal a systematic error but may reduce the effect of random errors. The error in the mean will be the average error of the trials.	Voltage across a resistor.
	Voltages: $V_1 = 2.99 \pm 0.01$ V, $V_2 = 3.02 \pm 0.01$ V and $V_3 = 3.01 \pm 0.01$ V
	Average voltage, V_{ave}:
Value: $\quad m_{ave} = \dfrac{m_1 + m_2 + m_3 + ... + m_n}{n}$	$V_{ave} = \dfrac{V_1 + V_2 + V_3}{3} = \dfrac{2.99 + 3.02 + 3.01}{3} = 3.007$ V
Error: $\quad \delta m_{ave} = \dfrac{\delta m_1 + \delta m_2 + \delta m_3 + ... + \delta m_n}{n}$	Error in voltage, $\delta \Delta V$:
	$\delta V_{ave} = \dfrac{0.01 + 0.01 + 0.01}{3} = 0.01$ V
Answer: $\quad m_{ave} \pm \delta m_{ave}$	Final answer:
	$\quad\quad V_{ave} = 3.01 \pm 0.01$ V

Multiples	Example
Measuring multiples of identical items or events will improve the precision of the measurement of a single item or event by reducing the size of the absolute error. The absolute error is divided by the number of multiple events or objects.	Time taken for a pendulum to complete a single oscillation.
	Time for 10 oscillations: $t_{10} = 20.1 \pm 0.1$ s
	Time for single oscillation, t_1:
Value: $\quad d_{single} = \dfrac{d_{multiple}}{\text{number of multiples}}$	$t_1 = \dfrac{t_{10}}{10} = \dfrac{20.1}{10} = 2.01$ s
	Error in time taken, $\delta \Delta t$:
Error: $\quad \delta d_{single} = \dfrac{\delta d_{multiple}}{\text{number of multiples}}$	$\delta t_1 = \dfrac{0.1}{10} = 0.01$ s
	Final answer:
Answer: $\quad d_{single} \pm \delta d_{single}$	$t_1 = 2.01 \pm 0.01$ s

Multiplication and division	Example
When two quantities are combined by multiplication or division, the error in the calculated value is determined by adding the relative errors.	The spring constant of a spring is determined by measuring the force and extension.
	Values: $F = 13.0 \pm 0.5$ N and $x = 0.25 \pm 0.01$ m
Value: $\quad F = m \times a$	Spring constant, k:
	$k = F/x = 13.0/0.25 = 52$ N m^{-1}
Error: $\quad \dfrac{\delta F}{F} = \dfrac{\delta m}{m} + \dfrac{\delta a}{a}$ hence	Error in spring constant, δk:
$\quad\quad \delta F = F \times \left(\dfrac{\delta m}{m} + \dfrac{\delta a}{a}\right)$	$\dfrac{\delta k}{k} = \dfrac{\delta F}{F} + \dfrac{\delta x}{x}$ so $\delta k = 52\left(\dfrac{0.5}{13.0} + \dfrac{0.01}{0.25}\right) = 4.08$ Nm^{-1}
	Final answer:
Answer: $\quad F \pm \delta F$	$k = 52 \pm 4$ N m^{-1}

Powers	Example
Non-linear relationships require the results of the experiment to be processed. Value: $T^n = T \times T \times T \times ... \times T_n$ Error: $\dfrac{\delta T^n}{T^n} = \dfrac{\delta T}{T} + \dfrac{\delta T}{T} + \dfrac{\delta T}{T} + ... + \dfrac{\delta T}{T_n}$ hence $\dfrac{\delta T^n}{T^n} = |n| \dfrac{\delta T}{T}$ hence $\delta T^n = T^n \times |n| \dfrac{\delta T}{T}$ (the notation $|n|$ means that the value is always positive) Answer: $T^n \pm \delta T^n$ Note: Always write roots or inverse relationships as powers.	The gravitational force between two large masses is given by the relationship $F = \dfrac{GMm}{r^2}$. Processing r to find $\dfrac{1}{r^2}$ and its error. Values: $r = 5.3 \pm 0.2$ m Processing $\dfrac{1}{r^2} = r^{-2}$: $r^{-2} = 5.3^{-2} = 0.036$ m^{-2} Error in r^{-2}: $\dfrac{\delta r^{-2}}{r^{-2}} = |-2| \dfrac{\delta r}{r}$ so $\delta r^{-2} = 0.036 \times |-2| \dfrac{0.2}{5.3}$ hence $\delta r^{-2} = 0.003$ m^2 Final answer: $r^{-2} = 0.036 \pm 0.003$ m^{-2}

Worked example: Combining errors

A ball of mass 0.65 ± 0.01 kg is thrown upwards at a speed of 4.00 ± 0.05 m s^{-1} from a height 3.33 ± 0.02 m. Determine the total energy of the ball (you may ignore internal energy) as it leaves the thrower's hand and determine the uncertainty in your answer. Explain whether this uncertainty is significant.

Take $g = 9.807 \pm 0.003$ m s^{-2}

Solution

The final answer will involve both gravitational and kinetic energy, which contain different quantities and different mathematical processes. Don't try to do the calculation of the value or the uncertainty in a single step, instead determine the value and uncertainty in each type of energy separately, then combine them to find the uncertainty in the total energy.

Given $m = 0.65 \pm 0.01$ kg, $v = 4.00 \pm 0.05$ m s^{-1}, $h = 3.33 \pm 0.02$ m, $g = 9.807 \pm 0.003$ m s^{-2}

Unknown $E_T = ?$

Equations $E_p = mgh$ and $E_K = \dfrac{1}{2}mv^2$ and $E_T = E_p + E_K$

Substitute   **Potential energy:** $E_p = mgh = 0.65 \times 9.807 \times 3.33 = 21.227$ J

Error: Multiplying three different quantities so combine using relative errors.

$\dfrac{\delta E_p}{E_p} = \dfrac{\delta m}{m} + \dfrac{\delta g}{g} + \dfrac{\delta h}{h}$ so $\dfrac{\delta E_p}{21.227} = \dfrac{0.01}{0.65} + \dfrac{0.003}{9.807} + \dfrac{0.02}{3.33}$ hence

$\delta E_p = 21.227 \times 0.0217 = 0.461$ J

Ans: $E_p = 21.2 \pm 0.5$ J

Kinetic energy: $E_K = \frac{1}{2}mv^2 = \frac{1}{2} \times 0.65 \times 4.00^2 = 5.2$ J

Error: Multiplying two different quantities so combine using relative errors. The number $\frac{1}{2}$ has no error associated with it so can be ignored for the error calculation.

$\frac{\delta E_k}{E_k} = \frac{\delta m}{m} + |2| \times \frac{\delta v}{v}$ so $\frac{\delta E_k}{5.2} = \frac{0.01}{0.65} + |2| \times \frac{0.05}{4.00}$ hence

$\delta E_k = 5.2 \times 0.04038 = 0.21$ J

Ans: $E_k = 5.2 \pm 0.2$ J

TOTAL energy: $E_T = E_p + E_K = 21.227 + 5.2 = 26.427$ J

Error: Adding quantities so add absolute errors.

$\delta E_T = \delta E_p + \delta E_k = 0.461 + 0.21 = 0.671$

Ans: $E_T = 26.4 \pm 0.7$ J

Solve

$\delta\% = \frac{0.671}{26.4} \times 100\% = 2.54\%$

The percentage error in the final answer is less than 5%, indicating that the uncertainty in the total energy is not significant.

Exercise 2A

1   Consider the following measurements and identify possible sources of error that will affect the accuracy and precision of the reading. Suggest techniques that could be used to overcome significant errors.

   a   Measuring the width of a sheet of paper using a thick wooden metre ruler.

   _____

   _____

   _____

   _____

   _____

   _____

   b   Measuring the time period of a mass oscillating (bouncing up and down) on a spring.

   _____

   _____

   _____

   _____

   _____

   _____

   _____

    c   Measuring the mass of a ball bearing (small metal ball).

_____

_____

_____

_____

_____

_____

2   A long narrow box is measured by placing it on top of a ruler with a scale divided into millimetres. The width of the box is measured by placing it on the ruler and the edges are found to be at 24.5 mm and 42.0 mm. The box is then turned so that its length can be measured, and the ends are found at 5.0 mm and 67.0 mm.

    a   Show that the width of the box is 18 ± 1 mm.

_____

_____

_____

    b   Show that the length of the box is 62 ± 1 mm.

_____

_____

_____

    c   Compare the percentage error in each measurement and comment on whether the uncertainties are significant.

_____

_____

_____

_____

_____

    d   Calculate the area of one face of the box and determine the uncertainty in the area by combining the relative errors.

_____

_____

_____

_____

    e   Calculate the percentage error in the area and compare the answer to the percentage errors in the length and width.

_____

_____

 ISBN: 9780170368179

3   Sam is sitting on a swing that is 2.800 ± 0.005 m long from the top to the centre of mass of Sam and the seat. Once the seat is swinging back and forth, Sam times how long it takes to complete 15 oscillations (swing back and forth). He repeats the measurement five times and records the results in the table shown below.

Time for 15 oscillations (s) $\pm$ 0.1 s					$t_{average}$ for 15 oscillations (s)	δt for 15 oscillations (s)	$t_{average}$ for 1 oscillation (s)	δt for 1 oscillation (s)
Trial 1	Trial 2	Trial 3	Trial 4	Trial 5				
50.8	50.2	50.0	47.0	50.5	50.4	$\pm$ 0.4	3.36	$\pm$ 0.03

a   Circle the anomalous reading in the table.

b   Show that the average time taken to complete 15 oscillations is 50.4 s.

_____

c   Show that the uncertainty in the time for 15 oscillations is $\pm$ 0.4 s by using the 'half range' method.

_____

d   Show that the time taken for a single oscillation is 3.36 ± 0.03s.

_____

_____

e   Given that the average time for a single oscillation is 3.36 s, suggest what went wrong during trial 4.

_____

_____

f   (OPTIONAL) Determine the standard deviation (see Appendix 2a) in the average time for 15 oscillations, then calculate the uncertainty in 1 oscillation.

_____

_____

The length of the swing and the time taken are related by the equation $T = 2\pi \sqrt{\dfrac{l}{g}}$.

g   Rearrange the equation to find the gravitational field strength, g, and determine the uncertainty in the final answer.

_____

_____

_____

_____

_____

_____

h   Calculate the percentage error and comment on the reliability of the calculated value for g.

_____

_____

4 A car travelling with an initial speed of $v_i = 13.6 \pm 0.8$ m s^{-1} accelerates in a straight line at $a = 1.2 \pm 0.2$ m s^{-2} over a distance $d = 192 \pm 5$ m.

a Use the equation $v_f^2 = v_i^2 + 2ad$ to determine the final speed, v_f, of the car.

_____

_____

_____

b Calculate the uncertainty in the final speed.

_____

_____

_____

_____

_____

_____

_____

_____

_____

_____

_____

_____

c Calculate the percentage error. Comment on the reliability of the calculated value and identify the main source of error.

_____

_____

_____

_____

_____

 ISBN: 9780170368179

Data processing

Results table

A results table must have sufficient columns to handle all the data, the uncertainties and any processing that is required for averages, multiple readings and processing the x-axis variable.

Independent variable (unit) Estimated uncertainty	Dependent variable (unit) Estimated uncertainty							If multiples have been recorded		Processed x-axis variable (processed unit)	
	Trial 1	Trial 2	Trial 3	Trial 4	Trial 5	Average	Uncertainty	Average for 1	Uncertainty for 1	Processed value	Processed uncertainty

Graphs — introduction

Line graphs are always* drawn with the dependent variable on the y-axis and the independent variable on the x-axis.

(*There are a few exceptions to the rule, for example, time-related quantities. When timing regular, repeated actions such as a swinging pendulum, the **time period** (T) (dependent variable) is a **discrete** event so can be plotted on the **y-axis**. But when timing a continuously changing event, such as the distance of a rolling ball, time (t) (dependent variable) is **continuous** so is plotted on the **x-axis** so that the gradient calculation reveals the rate of change.)

All graphs should include the following basic components:

- Title describing the two variables.
- Axis labelled with quantity and unit.
- Appropriate scale that allows all the points to be plotted accurately, and takes up the majority of the space on the graph paper. Do not use axis breaks (-∿-).
- Points plotted using crosses.
- Anomalous points should be identified by drawing a circle around them.
- A line of best fit (straight or smooth curve).

Title: 'Dependent vs Independent'

Graph with y-axis "Dependent variable (unit)" (0–12) and x-axis "Independent variable (unit)" (0–40), showing a curve with plotted crosses and one circled anomalous point.

If the graph is non-linear, it cannot be analysed until the data has been processed to produce a linear graph.

Processing non-linear graphs

There are four possible curved graph relationships that you will encounter (shown below) and each one has a characteristic shape. It is important that you can recognise which relationship is being studied.

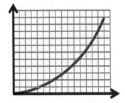 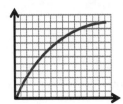 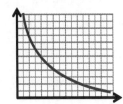 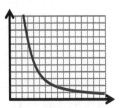

Once the type of relationship has been identified, the x-axis data (values and uncertainties) can be transformed. For example, for the relationship $a \propto b^2$, all the b values must be squared, and the uncertainty processed using the relationship $\frac{\delta b^2}{b^2} = |2| \frac{\delta b}{b}$.

A new graph using the processed values on the *x*-axis can now be plotted and should be linear.

Error bars and error lines

Once a linear graph has been generated, additional information can be added to enable further analysis of the data.

- **Error bars** drawn onto each point show the amount of uncertainty in that data, e.g. an error of ±1 N will go up and down by 1 N from the plotted point. If the error bars are so small that they won't show up on the graph, then state this underneath the graph.
- **Error lines** drawn in using the extreme error bars on the plotted points at either end.
  - Maximum error line — steepest possible gradient taken from the error bars.
  - Minimum error line — least steep gradient taken from the error bars.
  - The error line that has the greatest difference to the line of best fit should be used to generate the error in the gradient and intercept (see below).

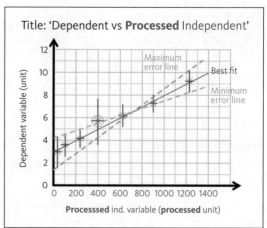

Title: 'Dependent vs **Processed** Independent'

Analysing linear graphs

The gradient and intercept of the line of best fit and the most significant error line can now be found from the processed graph. The uncertainty in the gradient and intercept can then be determined and the equation of the line presented.

Gradient, *m*

- The gradient of a straight line is a constant, so is often referred to as the constant of proportionality.
- Draw a triangle on the graph to determine the change in *y* and the change in *x*. (The triangle should be more than ⅔ the length of the line of best fit, and touch the line between plotted points.)
- The gradient, *m*, of each line is calculated by the equation:

$$m_b = \frac{\Delta y_b \ (y\,\text{unit})}{\Delta x_b \ (x\,\text{unit})} \quad \text{and} \quad m_e = \frac{\Delta y_e \ (y\,\text{unit})}{\Delta x_e \ (x\,\text{unit})}$$

- The unit of the gradient will be $\frac{(y\,\text{unit})}{(x\,\text{unit})}$.

- The uncertainty in the gradient: $\delta m = |m_b - m_e|$

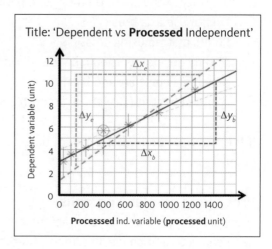

Title: 'Dependent vs **Processed** Independent'

Intercept, *c*

- The line of best fit and error line intercepts, c_b and c_e, are the points at which each line crosses the *y*-axis at *x* = 0.
- The unit of the intercepts will be the unit of the *y*-axis.
- The uncertainty in the intercept: $\delta c = |c_b - c_e|$

 ISBN: 9780170368179

Equation of the line

- The equation of a straight line is given by the formula: $y = (m \pm \delta m)x + (c \pm \delta c)$. The gradient, intercept, dependent and independent variables must be substituted into the formula along with their units.
- The percentage error is a good indicator of the reliability of the experiment. $\delta_{\%} m = \dfrac{\delta m}{m} \times 100\%$

Interpreting

Having completed the experiment and processed the data, it is essential that a conclusion is written which covers the following points.

Link the relationship to the aim and physics ideas

- Link the final result and any derived quantities with the aim of the experiment and known physics theories.
- Compare the final result with known physics theories/relationships/quantities and determine the validity of the experiment and whether further work needs to be carried out.
- Discuss any limitations to the theory's applicability. For example, the mathematical formula for a simple pendulum is derived using the small angle approximation so it is only valid when the angle through which the pendulum is displaced is kept below 10°.

Evaluate the experimental technique (if not previously discussed in the method)

- Discuss any limitations to the experiment or the apparatus used.
- Discuss the other variable(s) that could have significantly affected the results, and how they could have changed the results.
- Discuss any difficulties that were encountered and how they were overcome.

Evaluate the data

- Discuss inaccurate or anomalous data and suggest what may have caused it.
- Discuss any unexpected outcomes of the processing of the results and suggest how they could have been caused and the effect they had on the validity of the conclusion.
- Discuss the line of best fit, the gradient and the intercept compared to known physics relationships.

Exercise 2B

1 Ruby uses a spring to project a glider along an air track. The spring provides a constant change of momentum of the glider. She changes the mass of the glider and measures the speed for each mass using an electronic sensor.

a Identify any anomalous values and calculate the average (mean) and uncertainty for each data set. Write sample calculations below for the first line of the table to show how the values were calculated.

Mass (kg) (± 0.001 kg division error)	Speed (ms⁻¹) (± 0.5 ms⁻¹ division error)							Processed x-axis variable (processed unit)	
	Trial 1	Trial 2	Trial 3	Trial 4	Trial 5	Average	Uncertainty	Processed value	Processed uncertainty
0.050	7.7	6.8	7.0	6.8	7.2				
0.100	3.4	3.9	3.2	4.1	2.8				
0.150	1.8	2.0	2.2	2.2	3.1				
0.200	1.7	1.7	2.4	2.1	2.5				
0.250	1.6	0.5	1.1	1.4	1.7				
0.300	0.8	1.3	1.9	0.8	1.5				
0.350	0.9	0.9	0.9	1.4	0.5				

Sample calculation of the average speed: _____

_____

Sample calculation of the uncertainty in the average speed: _____

_____

b Draw a graph of speed (y) against mass (x) using the data above, and draw on the line of best fit.

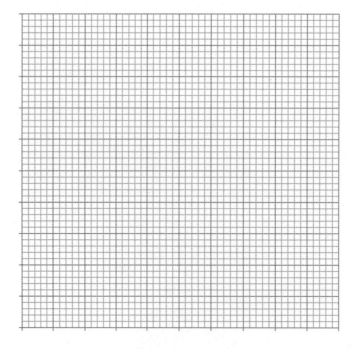

c   State the relationship based upon the shape of the graph.

_____

d   Write sample calculations below for the first line of the table to show how the *x*-axis (mass) values are processed, then complete the last two columns of the table by processing the *x*-axis data and the uncertainties. Include an appropriate quantity and unit in the table header.

Sample calculation of the processed mass: _____

_____

Sample calculation of the uncertainty in the processed mass: _____

_____

e   Plot a second graph below using your processed data. Draw the line of best fit and the maximum and minimum error lines.

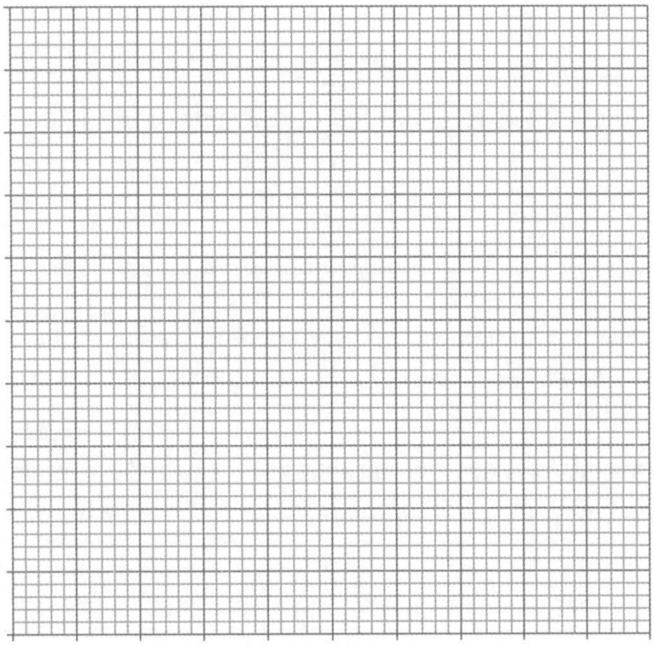

f   Determine the gradient of the graph and the intercept and state the equation of the line. Provide units with all your values.

_____

_____

_____

_____

g Decide which error line is most significantly different to the best-fit line and determine its gradient and intercept. Provide units with all your values.

_____

_____

_____

h Determine the error in the gradient and the error in the intercept.

_____

_____

_____

i Present the equation of the line of best fit with its uncertainties and units.

_____

_____

j Compare the equation of the line to the theoretical equation $p = mv$ and hence determine the momentum of the glider and the uncertainty in the momentum.

_____

_____

k Calculate the percentage error in the momentum and comment on the reliability of the data.

_____

_____

2 Vincent spins a small mass around his head. He keeps the radius of the circle constant at 70.0 ± 0.1 cm but increases the speed and measures the size of the force using a newton meter. The data is shown below.

a Identify any anomalous values and calculate the mean and uncertainty for each data set. Write sample calculations below for the first line of the table to show how the values were calculated.

Speed (m s⁻¹) (± 0.1 m s⁻¹ division error)	Force (N) (± 0.2 N parallax error)							Processed x-axis variable (processed unit)	
	Trial 1	Trial 2	Trial 3	Trial 4	Trial 5	Average	Uncertainty	Processed value	Processed uncertainty
3.0	0.6	0.4	0.7	0.5	1.0				
4.0	1.1	1.2	1.2	1.2	1.0				
5.0	1.6	1.4	1.7	1.5	1.1				
6.0	1.9	1.9	2.3	2.2	1.8				
7.0	2.8	2.6	2.9	2.3	3.0				
8.0	3.7	3.4	3.7	3.9	4.0				
9.0	4.5	4.7	4.6	4.7	4.7				

 ISBN: 9780170368179

Sample calculation of the average force: _____

_____

Sample calculation of the uncertainty in the average force: _____

_____

b Draw a graph of force (*y*) against speed (*x*) using the data above, and draw on the line of best fit.

c State the relationship based upon the shape of the graph.

d Write sample calculations below for the first line of the table to show how the *x*-axis (speed) values are processed, then complete the last two columns of the table by processing the *x*-axis data and the uncertainties. Include an appropriate quantity and unit in the table header.

Sample calculation of the processed speed: _____

_____

Sample calculation of the uncertainty in the processed speed: _____

_____

e Plot a second graph below using your processed data. Draw the line of best fit and the maximum and minimum error lines.

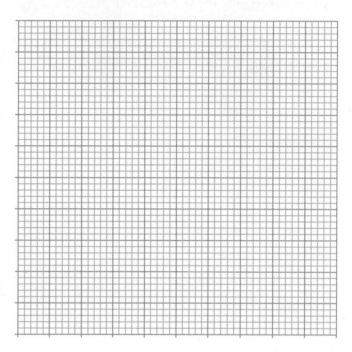

f Determine the gradient of the graph and the intercept and state the equation of the line. Provide units with all your values.

_____

_____

_____

g Decide which error line is most significantly different to the best-fit line and determine its gradient and intercept. Provide units with all your values.

_____

_____

_____

h Determine the error in the gradient and the error in the intercept.

_____

_____

_____

i Present the equation of the line of best fit with its uncertainties and units.

_____

_____

j Compare the equation of the line to the theoretical equation $F = \dfrac{mv^2}{r}$ and hence determine the mass of the object and the uncertainty in the mass.

_____

_____

_____

_____

_____

k Calculate the percentage error in the mass and comment on the reliability of the data.

_____

_____

_____

3 Debbie wants to investigate the relationship between the time period of a pendulum oscillating (swinging) on a string and the length of the string. Her data is shown below.

a Identify any anomalous values and calculate the mean and uncertainty for each data set. Write sample calculations below for the first line of the table to show how the values were calculated.

Length (m) (± 0.005 m division error)	Time period for 10 swings (s) (± 0.5 s time reaction error)							Processed x-axis variable (processed unit)	
	Trial 1	Trial 2	Trial 3	Trial 4	Trial 5	Average for 10	Uncertainty in t for 10	Processed value	Processed uncertainty
0.200	8.6	9.5	9.4	8.0	8.9				
0.400	12.9	11.8	11.9	12.8	12.7				
0.600	14.9	15.2	16.2	15.4	16.0				
0.800	18.3	18.7	18.9	18.3	17.7				
1.000	19.8	20.8	20.7	19.9	20.2				
1.200	21.7	21.2	22.1	22.0	22.6				
1.400	23.1	24.2	22.2	23.6	24.0				

Sample calculation of the average time for **1 swing**: _____

_____

Sample calculation of the uncertainty in the average time for **1 swing**: _____

_____

b Draw a graph of time for 1 swing (y) against length (x) using the data above, and draw on the line of best fit.

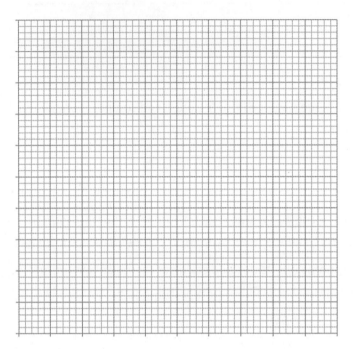

c State the relationship based upon the shape of the graph.

_____

d Write sample calculations below for the first line of the table to show how the x-axis values are processed, then complete the last two columns of the table by processing the x-axis (length) data and the uncertainties. Include an appropriate quantity and unit in the table header.

Sample calculation of the processed length: _____

_____

Sample calculation of the uncertainty in the processed length: _____

_____

e Plot a second graph below using your processed data. Draw the line of best fit and the maximum and minimum error lines.

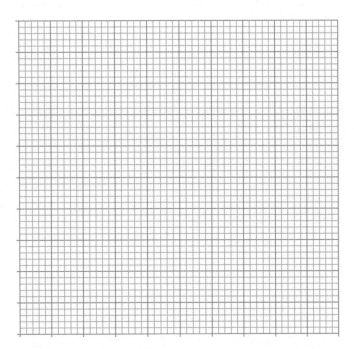

f Determine the gradient of the graph and the intercept and state the equation of the line. Provide units with all your values.

_____

_____

_____

g Decide which error line is most significantly different to the best-fit line and determine its gradient and intercept. Provide units with all your values.

_____

_____

_____

h Determine the error in the gradient and the error in the intercept.

_____

_____

_____

i Present the equation of the line of best fit with its uncertainties and units.

_____

_____

j Compare the equation of the line to the theoretical equation $T = 2\pi\sqrt{\dfrac{l}{g}}$ and hence determine a value for the acceleration due to gravity, g.

_____

_____

_____

_____

_____

k Calculate the percentage error in the acceleration due to gravity and compare your answer to the accepted value of 9.81 m s⁻².

_____

_____

_____

_____

_____

4 Nehe wants to investigate how the strength of the electric field around a small charged ball changes with distance from the ball. The data is shown below.

   a Identify any anomalous values and calculate the mean and uncertainty for each data set. Write sample calculations below for the first line of the table to show how the values were calculated.

Distance, d (m) ($\pm$ 0.002 m division error)	Electric field strength, E (V m⁻¹) ($\pm$ 0.05 V m⁻¹ time reaction error)							Processed x-axis variable (processed unit)	
	Trial 1	Trial 2	Trial 3	Trial 4	Trial 5	Average	Uncertainty	Processed value	Processed uncertainty
0.050	16.09	16.09	16.05	16.11	16.11				
0.100	3.99	4.05	4.09	3.99	4.03				
0.150	1.73	1.79	1.81	1.77	1.79				
0.200	1.07	1.05	0.95	1.05	0.99				
0.250	0.62	0.68	0.64	0.64	0.66				
0.300	0.51	0.43	0.49	0.43	0.45				
0.350	0.31	0.29	0.29	0.35	0.33				

Sample calculation of the average electric field strength: _____

_____

Sample calculation of the uncertainty in the average electric field strength:

_____

_____

b Draw a graph of electric field strength (*y*) against distance (*x*) using the data above, and draw on the line of best fit.

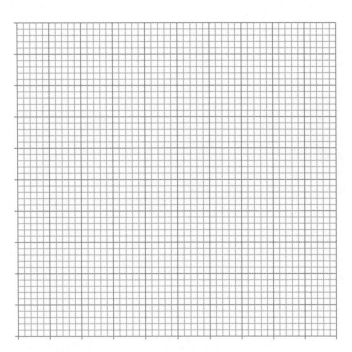

c State the relationship based upon the shape of the graph.

_____

d Write sample calculations below for the first line of the table to show how the *x*-axis (distance) values are processed, then complete the last two columns of the table by processing the *x*-axis data and the uncertainties. Include an appropriate quantity and unit in the table header.

Sample calculation of the processed distance: _____

_____

Sample calculation of the uncertainty in the processed distance: _____

_____

e Plot a second graph below using your processed data. Draw the line of best fit and the maximum and minimum error lines.

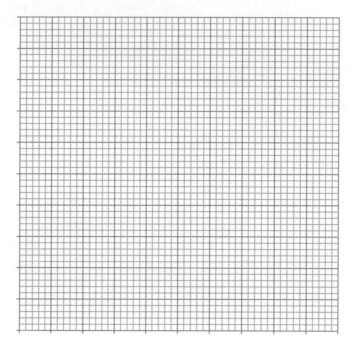

f Determine the gradient of the graph and the intercept and state the equation of the line. Provide units with all your values.

_____

_____

_____

g Decide which error line is most significantly different to the best-fit line and determine its gradient and intercept. Provide units with all your values.

_____

_____

_____

h Determine the error in the gradient and the error in the intercept.

_____

_____

_____

i Present the equation of the line of best fit with its uncertainties and units.

_____

_____

_____

 ISBN: 9780170368179

j Compare the equation of the line to the theoretical equation $E = \dfrac{kq}{d^2}$. Given that Coulomb's constant $k = 8.987 \pm 0.0005 \times 10^9$ V m C^{-1}, determine the size of the charge on the ball and the uncertainty in the charge.

_____

_____

_____

_____

_____

k Calculate the percentage error in the charge and comment on the reliability of the data.

_____

_____

_____

Exercise 2C

The following sample reports contain errors and/or omissions. Read each one and assess them using the mark schedule at the end of each report. You should annotate each report to identify any issues.

Student report 1: Time period of a bouncing mole mobile

Aim
To determine the relationship between the mass of a toy and the time taken for it to complete a single oscillation for a baby's decorative mobile when attached to a spring with a spring constant of $k = 3.00 \pm 0.04$ N m^{-1}.

Hypothesis
When pulled down and released, the suspended mass will oscillate in simple harmonic motion (SHM), which is given by the formula $T = 2\pi\sqrt{\dfrac{m}{k}}$. I predict that increasing the mass will increase the time.

Method
- The independent variable will be the mass hanging from the spring. It will range from 50 g to 350 g in 50 g intervals. The maximum mass is limited by the length of the spring when suspended over the edge of the lab bench (112 cm long). This meant it could still oscillate without hitting the ground.
- The dependent variable will be the time taken to complete one oscillation (the time period). I will measure the time taken for the mass to travel from the bottom to the top and back again.
- I will repeat the measurement five times for each mass and calculate the average time.
- During the experiment I will always displace the mass the same small distance from the equilibrium position to ensure the amplitude of the oscillation is kept constant. This will mean that air resistance is kept to a minimum for each trial.

Diagram of the experimental set-up

Techniques to improve accuracy

- I estimate that my reaction time error will be ± 0.2 s. I will measure the dependent variable five times and calculate the averages. This will help to reduce the effect of random errors due to my reaction time errors. I will also ensure that I am eye-level with the bottom of the oscillation to reduce parallax error when trying to judge exactly when the mass reaches the bottom.

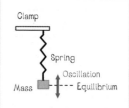

Results

The time taken, t, for a mass, m, to complete a single oscillation is recorded in the table below.

Mass (g)	Time period for an oscillation (s) (± 0.2 s reaction time error)							Mass$^{1/2}$ (g$^{1/2}$)
	Trial 1	Trial 2	Trial 3	Trial 4	Trial 5	Average t_{ave}	Uncertainty $\delta t_{1/2}$	
50	0.7	0.8	1.0	0.6	0.7	0.8	0.2	7.07
100	1.2	1.3	1.2	1.1	1.3	1.2	0.1	10.00
150	1.3	1.3	1.4	1.4	1.4	1.4	0.1	12.25
200	1.7	1.6	1.6	1.7	1.8	1.7	0.1	14.14
250	1.7	2.0	1.9	1.7	1.8	1.8	0.2	15.81
300	1.9	2.0	2.0	2.1	1.9	2.0	0.1	17.32
350	2.1	2.1	2.2	2.0	2.3	2.1	0.2	18.71

The averages were calculated as follows:

$$t_{ave} = \frac{0.7 + 0.8 + 1.0 + 0.6 + 0.7}{5} = 0.8 \text{ s}$$

The uncertainty in the time was calculated using half the range as follows:

$$\delta t_{1/2} = \frac{1.0 - 0.6}{2} = 0.2 \text{ s (1 sf)}$$

A graph of average time against mass reveals the relationship as:

Time is proportional to the square root of the mass.

My prediction that both increased was correct but it did not describe the exact relationship. I will now square root all the masses and draw a graph of time against the square root of mass.

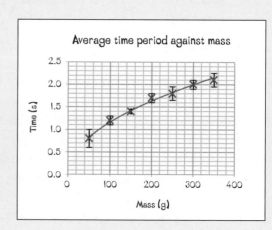

Line of best fit
Gradient:

$$m_b = \frac{\Delta T_b}{\Delta m_b^{1/2}} = \frac{2.05 - 1.00}{18.00 - 8.50} = 0.11 \text{ s g}^{-1/2}$$

Intercept: $c_b = 0.05$ s

Error line
Gradient:

$$m_e = \frac{\Delta T_e}{\Delta m_e^{1/2}} = \frac{1.80 - 1.05}{17.00 - 8.00} = 0.08 \text{ s g}^{-1/2}$$

Intercept: $c_e = 0.4$ s

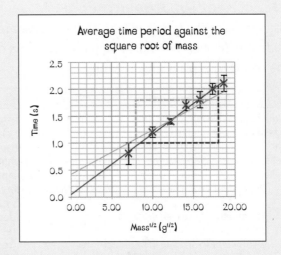

Average time period against the square root of mass

Uncertainties
Gradient: $\delta m_b = |0.11 - 0.08| = \pm 0.03 \text{ s g}^{-1/2}$
Intercept: $\delta c_b = |0.05 - 0.4| = \pm 0.4$ s

Equation of the relationship
Equation: $T = (0.11 \pm 0.03)m^{1/2} + (0.0 \pm 0.4)$
　　　　　(s)　　　(s g$^{-1/2}$)　(g$^{1/2}$)　　　(s)

Conclusion
The experimental equation of the relationship is: $T = (0.11 \pm 0.03)\sqrt{m} + (0.0 \pm 0.4)$
　　　　　　　　　　　　　　　　　　　　　　　　　(s)　　　(s g$^{-1/2}$)　(g$^{1/2}$)　　(s)

which shows that $T \propto \sqrt{m}$ and agrees with the theoretical equation of: $T = 2\pi \sqrt{\dfrac{m}{k}}$

The percentage error in my gradient is: $\delta_\% = \dfrac{0.03}{0.11} \times 100 = 27\%$, which is very large and

when I put a mass of 50 g into both equations gives: $T_{exp} = 0.11 \times \sqrt{50} = 0.78$ s and

$T_{theory} = 2\pi \sqrt{\dfrac{50}{3.0}} = 25.7$ s. These are very different values and suggests that I have done the

experiment very badly.

Evaluation
It was very difficult to measure the time of each oscillation accurately, particularly the small masses because they took such a short amount of time. To overcome this issue I could have timed several oscillations and divided by the number of oscillations. By taking multiples this would have made it easier to measure.

My experiment probably didn't model a child's toy very well as it was made of brass and very heavy, which would not be safe. A toy for a child would be much lighter and softer and be bigger in size so have a lot more air resistance, which would affect the speed it moves up and down and cause the time to take longer, which is why my value is so different to the theoretical value.

An unexpected outcome of the experiment occurred when I suspended a 250 g mass from the spring. When I set it oscillating, the amplitude of the oscillation got gradually smaller until it stopped bouncing and started swinging from side to side, then it gradually stopped swinging and started bouncing again. This made it very hard to measure the time period for this mass.

Mark sheet for Student report 1

Use the following schedule to mark the student report and determine the appropriate grade.

Identify (circle/underline/annotate) any errors or omissions in the report and write a brief note on the report to explain what is wrong.

	Achieved	✓	Merit	✓	Excellence	✓

Planning and Gathering	**A1** Collect data relevant to the aim based on the manipulation of the independent variable over a reasonable range and number of values:
	• Dependent and independent values measured.
	• Reasonable range (see graph to determine if the range is sufficient to show the correct shape).
	• Five or more different values of the independent variable tested.
		☐	**M1** Describe the control of other variable(s) that could significantly affect the results.
			• Example 1 (required)
			• Example 2
			M2 Use technique(s) to improve the accuracy of the measured values of:
			• the dependent variable
			  – repeats and averages (required)
			• and independent variable
			  – parallax error
			  – zero error.
				☐
				
				☐
					E1 Discuss the other variable(s) that could be changed and would significantly affect the results, and what effect they would have.	☐

Processing	**A2** Appropriate estimate of uncertainties in the raw data for **EITHER**: • the independent variable • the dependent variable. **A3** Draw an **acceptable** linear graph from the processed data that shows the relationship between the independent and dependent variables: • Axis labels and units. • Linear scales. • Majority of points plotted correctly. • Line of best fit. • Error line based on data points. **A4** Use the graph to determine: • the gradient • the intercept • of the line of best fit. **AND** state the: • equation of the relationship (required) • the value of the physics quantity (if applicable).	☐ ☐ ☐	**M3** Processes uncertainties in **EITHER**: • the independent variable • the dependent variable. **M4** Draw an **appropriate** linear graph from the processed data that shows the relationship between the independent and dependent variables as for Achieved **PLUS**: • Draw error bars, and/or. • Draw a maximum or minimum error line based on the error bars. **M5** Use the **graph** to determine: • the gradient (required) and • the intercept • of the error line. **AND** determine the **uncertainty** in: • the gradient (required) • the intercept • of the line of best fit. **AND** state the: • equation of the **relationship** AND • its **uncertainty**.	☐ ☐ ☐		
Interpreting	**A5** Make a comparison with the physics theory and: • the relationship AND/OR • the quantity obtained from the experimental data. Candidates could compare: • the type of relationship found experimentally with that which is expected • a derived quantity with a theoretical value • an experimental gradient with the theoretical gradient • the experimental intercept with the expected intercept.	☐	**M6** Make a quantitative comparison between the physics theory and: • the relationship AND/OR • the quantity obtained from the experimental data **WHICH** • includes consideration of the uncertainties. Candidates could determine: • the difference between calculated values derived from the theoretical and experiment relationships • the difference between a derived quantity and the theoretical value.	☐	**E2** Discuss the limitations to the application of physics theories: • to the practical situation and/or • at extreme values of the independent variable. • Examples 1 and 2 **E3** Discuss any unexpected outcomes of the processing of the results and a suggestion of how they could have been caused and the effect they had on the validity of the conclusion.	☐ ☐
	Achieved: All points correct		**Merit:** All points correct		**Excellence:** 2 good or 3 reasonable points	
					Overall Grade	

 ISBN: 9780170368179

Student report 2: Time period of a bouncing mole mobile

Aim

To determine the relationship between the mass of a toy and the time taken for it to complete a single oscillation for a baby's decorative mobile when attached to a spring with a spring constant of $k = 3.00 \pm 0.04$ N m^{-1}.

Hypothesis

When a mass is suspended from a spring, the weight force causes the spring to extend until the restoring force of the spring balances the weight. At this point the system (spring and mass) is in equilibrium. When pulled down and released, the suspended mass will oscillate about the equilibrium position in simple harmonic motion (SHM). The theoretical SHM relationship between the time period and the mass is given by the formula $T = 2\pi\sqrt{\dfrac{m}{k}}$, which can be rearranged to $T = \dfrac{2\pi}{\sqrt{k}}\sqrt{m}$ showing that $T \propto \sqrt{m}$. I predict, therefore, that the experiment should reveal that the time is proportional to the square root of the mass.

Method

- The independent variable will be the mass hanging from the spring.
    - The 0.050 kg masses I will be using were checked on a balance and found to range from 0.047 up to 0.052 kg. I have therefore estimated the error in each mass to be $\pm$ 0.003 kg.
    - The masses will range from 0.050 $\pm$ 0.003 kg to 0.350 $\pm$ 0.02 kg in 0.050 kg intervals.
    - Preliminary testing revealed that significant resonance occurred when 0.250 kg was placed on the clamp stand causing the mass to stop moving up and down and instead swing sideways making it difficult to time. This occurs when the frequency of the oscillating mass matches the natural frequency of the clamp stand causing energy from the oscillating spring to be transferred to the clamp stand making it move. I will therefore not use 0.250 kg in my investigation.

- The time taken to complete one oscillation (the time period) is the dependent variable.
    - I estimate that my reaction time error will be $\pm$ 0.2 s. From preliminary tests I realised that this resulted in a very significant percentage error for smaller masses:

$$\frac{0.2}{0.8} \times 100\% = 25\%$$

    - I will therefore measure the time taken for the mass to complete 10 oscillations then divide this by 10 to find the time for a single oscillation. This reduces the percentage error to an acceptable size:

$$\frac{0.2}{8.1} \times 100\% = 2.5\%$$

    - I will repeat the measurement five times for each mass and calculate the average time. This will help to reduce the effect of random errors due to my reaction time errors.

- The following variables were controlled.
  - I will use the same spring, masses and measuring equipment throughout.
  - I will always release it from 1.0 cm below the equilibrium position. As each mass is added to the stack, the total surface area will increase, which will increase air resistance, however my preliminary work showed that this does not affect the time period. I will, however, keep the amplitude small as this reduces the speed at which the mass bounces reducing the air resistance effects and making it easier to judge when it reaches the bottom.

Diagram of the experimental set-up

Techniques to improve accuracy

- I checked the mass of my masses using scales. The scales had a tare button so I could avoid zero error.
- I will be measuring multiple oscillations and taking repeats and averages (see method).
- I will also avoid parallax error by getting eye-level with the bottom of the mass when I release and time it.

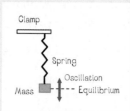

Results

Mass (g)	Time period for an oscillation (s) (±0.2 s reaction time error)							Time period for a single oscillation (s)		Mass$^{1/2}$ (kg$^{1/2}$)	δm (kg$^{1/2}$)
	Trial 1	Trial 2	Trial 3	Trial 4	Trial 5	Average t_{ave}	Uncertainty $\delta t_{1/2}$	Average $t_{1\,ave}$	Uncertainty $\delta t_{1/2}$ for 1		
0.050	8.2	8.3	8.1	8.0	8.0	8.1	0.2	0.81	0.02	0.22	0.01
0.100	11.6	11.4	11.4	11.5	11.3	11.4	0.2	1.14	0.02	0.316	0.009
0.150	14.2	14.3	14.2	13.8	14.1	14.1	0.3	1.41	0.03	0.39	0.01
0.200	16.3	16.2	16.1	16.4	16.1	16.2	0.2	1.62	0.02	0.45	0.01
0.300	20.0	19.7	19.7	20.1	19.8	19.9	0.2	1.99	0.02	0.55	0.02
0.350	21.5	21.6	21.3	21.2	21.5	21.4	0.2	2.14	0.02	0.59	0.02

The averages were calculated as follows:

$$\text{For 10 oscillations } t_{10\,ave} = \frac{8.2 + 8.3 + 8.1 + 8.0 + 8.0}{5} = 8.1 \text{ s so for 1 oscillation}$$

$$t_{1\,ave} = \frac{8.1}{10} = 0.81 \text{ s (2 sf)}$$

The uncertainty in the time was calculated using half the range as follows:

$$\text{For 10 oscillations } \delta t_{1/2} = \frac{8.3 - 8.0}{2} = 0.2 \text{ s so for 1 oscillation}$$

$$\delta t_{1/2} = \frac{0.2}{10} = \pm 0.02 \text{ s (1 sf)}$$

The processing of the mass to get the square root of the mass:

$$m^{1/2} = 0.050^{1/2} = 0.2236 = 0.22 \text{ kg}^{1/2} \text{ (2 sf)}$$

The uncertainty in the square root of the mass:

$$\frac{\delta m^{1/2}}{m^{1/2}} = \left|\frac{1}{2}\right| \frac{\delta m}{m} \text{ so } \delta m^{1/2} = \left|\frac{1}{2}\right| \times \frac{\delta m}{m^{1/2}} = \left|\frac{1}{2}\right| \times \frac{0.003}{0.050^{1/2}} = \pm 0.01 \text{ kg}^{1/2} \text{ (2 dp)}^*$$

$$^* \text{ to match the dp in the mass above.}$$

 ISBN: 9780170368179

A graph of average time against mass reveals the relationship as:

Time is proportional to the square root of the mass, which agrees with the theoretical relationship for SHM. I will now square root all the masses and draw a graph of time against the square root of mass.

Average time period against mass

Line of best fit
Gradient:

$$m_b = \frac{\Delta T_b}{\Delta m_b^{1/2}} = \frac{2.10 - 0.30}{0.58 - 0.08} = 3.6 \text{ s kg}^{-1/2}$$

Intercept: $c_b = 0.00$ s

Average time period against the square root of mass

Error line
Gradient:

$$m_e = \frac{\Delta T_e}{\Delta m_e^{1/2}} = \frac{2.10 - 0.40}{0.56 - 0.12} = 3.9 \text{ s kg}^{-1/2}$$

Intercept: $c_e = -0.5$ s

Uncertainties
Gradient: $\delta m_b = |3.6 - 3.9| = \pm0.3$ s kg$^{-1/2}$
Intercept: $\delta c_b = |0.00 - (-0.5)| = \pm0.5$ s

Equation of the relationship
Equation: $T = (3.6 \pm 0.3)m^{1/2} + (0.0 \pm 0.5)$
$\quad$ (s) $\quad$ (s kg$^{-1/2}$) (kg$^{1/2}$) $\quad$ (s)

Conclusion
The experimental equation of the relationship is: $T = (3.6 \pm 0.3)\sqrt{m} + (0.0 \pm 0.5)$
$\quad$ (s) $\quad$ (s kg$^{-1/2}$) (kg$^{1/2}$) $\quad$ (s)

which agrees with the theoretical equation of: $T = 2\pi\sqrt{\dfrac{m}{k}}$

The percentage error in my gradient is: $\delta_\% = \dfrac{0.3}{3.6} \times 100 = 8.3\%$, which is reasonable and shows that the experimental technique was reliable.

Equating the equations reveals that: $3.6 \pm 0.3 = \dfrac{2\pi}{\sqrt{k}}$ so $\sqrt{k} = 1.745 \pm 0.15$ hence

$k = 3.05$ kg s^{-2} and the uncertainty is: $\dfrac{\delta k}{k} = |2|\dfrac{\delta\sqrt{k}}{\sqrt{k}}$ so $\delta k = 2 \times 1.745 \times 0.15 = \pm0.5$ kg s^{-2}

The experimental value for the spring constant is: $k = 3.0 \pm 0.5$ kg s^{-2}, which is in close agreement with provided value of $k = 3.00 \pm 0.04$ N m^{-1} and confirms the validity of the process used to determine the answer.

The line of best fit also passes through the origin as expected.

Evaluation

The similarity between my repeat time measurements, use of multiples and the close proximity of the data points to the line of best fit all reveal the accuracy to which the experiment has been conducted. This is reflected in the small 8.3% error in the gradient and similarity between the experimental and provided values of k.

The experiment confirmed the relationship between mass and spring constant and could be used to determine the ideal mass for a toy to provide the slow oscillation to entertain a baby without being too massive to cause injury. From my results, a long time period required a larger mass, as, $T \alpha \sqrt{m}$ which could be dangerous, so the toy might require a weaker spring with a lower k value as, $T \alpha \dfrac{1}{\sqrt{k}}$ so that smaller masses could be used and still have a long oscillation time.

An unexpected outcome of the experiment was the 6% error in the brass slotted masses that I used. This contributed a much more significant uncertainty in the line of best fit than the uncertainty in the time, which was very small due to the use of multiples. To reduce this error, I could have measured each mass individually and added the masses together as I increased the mass. This would mean I increased the mass in irregular intervals but would have reduced my uncertainty in the mass to ± 0.01, which is the uncertainty in the scales that I used to measure the mass.

 ISBN: 9780170368179

Mark sheet for Student report 2

Use the following schedule to mark the student report and determine the appropriate grade.

Identify (circle/underline/annotate) any errors or omissions in the report and write a brief note on the report to explain what is wrong.

	Achieved	✓	Merit	✓	Excellence	✓
Planning and Gathering	**A1** Collect data relevant to the aim based on the manipulation of the independent variable over a reasonable range and number of values: • Dependent and independent values measured. • Reasonable range (see graph to determine if the range is sufficient to show the correct shape). • Five or more different values of the independent variable tested.	☐	**M1** Describe the control of other variable(s) that could significantly affect the results. • Example 1 (required) • Example 2 **M2** Use technique(s) to improve the accuracy of the measured values of: • the dependent variable 　– repeats and averages (required) • and independent variable 　– parallax error 　– zero error.	☐ ☐	**E1** Discuss the other variable(s) that could be changed and would significantly affect the results, and what effect they would have.	☐
Processing	**A2** Appropriate estimate of uncertainties in the raw data for **EITHER**: • the independent variable • the dependent variable. **A3** Draw an **acceptable** linear graph from the processed data that shows the relationship between the independent and dependent variables: • Axis labels and units. • Linear scales. • Majority of points plotted correctly. • Line of best fit. • Error line based on data points. **A4** Use the graph to determine: • the gradient • the intercept • of the line of best fit. **AND** state the: • equation of the relationship (required) • the value of the physics quantity (if applicable).	☐ ☐ ☐	**M3** Processes uncertainties in **EITHER**: • the independent variable • the dependent variable. **M4** Draw an **appropriate** linear graph from the processed data that shows the relationship between the independent and dependent variables as for Achieved **PLUS**: • Draw error bars, and/or. • Draw a maximum or minimum error line based on the error bars. **M5** Use the **graph** to determine: • the gradient (required) and • the intercept • of the error line. **AND** determine the **uncertainty** in: • the gradient (required) • the intercept • of the line of best fit. **AND** state the: • equation of the **relationship** AND • its **uncertainty**.	☐ ☐ ☐		
Interpreting	**A5** Make a comparison with the physics theory and: • the relationship AND/OR • the quantity obtained from the experimental data. Candidates could compare: • the type of relationship found experimentally with that which is expected • a derived quantity with a theoretical value • an experimental gradient with the theoretical gradient • the experimental intercept with the expected intercept.	☐	**M6** Make a quantitative comparison between the physics theory and: • the relationship AND/OR • the quantity obtained from the experimental data **WHICH** • includes consideration of the uncertainties. Candidates could determine: • the difference between calculated values derived from the theoretical and experiment relationships • the difference between a derived quantity and the theoretical value.	☐ ☐	**E2** Discuss the limitations to the application of physics theories: • to the practical situation and/or • at extreme values of the independent variable. • Examples 1 and 2 **E3** Discuss any unexpected outcomes of the processing of the results and a suggestion of how they could have been caused and the effect they had on the validity of the conclusion.	☐ ☐
	Achieved: All points correct	☐	**Merit:** All points correct	☐	**Excellence:** 2 good or 3 reasonable points	☐
					Overall Grade	☐

3 Waves

3.0 Fundamental properties

Types of waves

A **progressive wave** or travelling wave is a method by which energy is transferred from one region to another without the transfer of matter between the regions. There are two types of wave:

- **Mechanical waves** are produced by a periodic disturbance (oscillation or vibration) in a material medium. The particles of the medium oscillate about fixed points and the energy is transmitted by subsequent particles vibrating. There is no real transfer of matter. Examples include: waves on strings or springs; water waves; sound waves; seismic waves.

- **Electromagnetic waves** are produced by disturbances in the electric and magnetic fields. No medium is necessary and they travel best through a vacuum. The electromagnetic (EM) spectrum is made up of: gamma rays, X-rays, ultraviolet, visible light, infrared, microwave and radio.

Types of wave motion

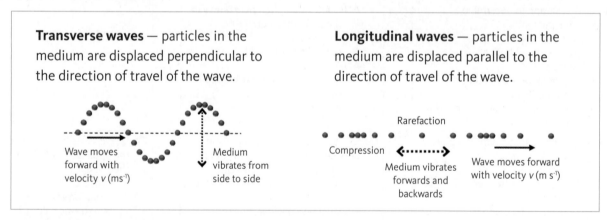

Transverse waves — particles in the medium are displaced perpendicular to the direction of travel of the wave.

Wave moves forward with velocity v (ms^{-1})

Medium vibrates from side to side

Longitudinal waves — particles in the medium are displaced parallel to the direction of travel of the wave.

Rarefaction

Compression

Medium vibrates forwards and backwards

Wave moves forward with velocity v (m s^{-1})

Representing waves

Both transverse and longitudinal waves can be represented as a sine wave with certain properties.

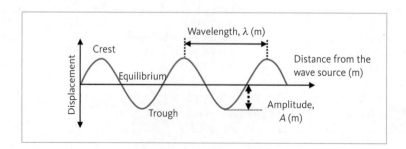

Wavelength, λ (m)

Crest

Equilibrium

Displacement

Distance from the wave source (m)

Trough

Amplitude, A (m)

 ISBN: 9780170368179

Wavelength (λ) (metre, m) — the distance between two points on a progressive wave with the same phase (e.g. crest and crest). (See below for the definition of phase.)

Amplitude (A) (metre, m) — the greatest displacement of the wave from its equilibrium position. The amplitude describes how much energy a wave transfers.

Period (T) (second, s) — the time taken for one complete wave to pass a point, or the time taken for a part of the wave to complete one oscillation.

Frequency (f) (hertz, Hz) — the number of complete waves produced each second, or the number of complete waves to pass a point each second, measured in hertz (Hz) or per second (s^{-1}). The frequency and time period are related by the relationship:

$$f = \frac{1}{T}$$

Velocity (v) (metre per second, m s^{-1}) — the velocity, frequency and wavelength of a wave are related by the formula:

$$v = f\lambda$$

Phase

The position of part of a wave in its cycle can be described in terms of its phase. Two points are in phase if their maximum and minimum values occur at the same instant, otherwise there is said to be a phase difference, which is measured in degrees or radians.

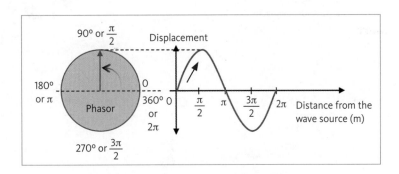

Wave fronts

Wave diagrams can also be drawn by considering the waves as viewed from above. The crest of each wave is represented by a solid line, called a **wave front**. Each wave front is one wavelength apart. (The troughs are sometimes shown as dotted lines in between the crests.)

Circular wave fronts are produced by point sources, for example a stone dropped into water.

Plane wave fronts are produced by straight sources, for example a stick dropped into water.

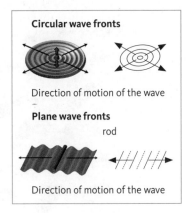

Diffraction

When waves pass through a gap, they spread out into the region either side of the gap, and as the width of the gap decreases, the amount of spreading increases.

When the **size of the gap is of a similar** size to the **wavelength**, the diffraction effects become very pronounced, producing semi-circular wave fronts.

As the size of the gap decreases, the amount of energy that passes through the gap decreases, reducing the amplitude of the wave, but the wavelength remains constant. As the energy becomes spread out due to diffraction, the amplitude decreases even more.

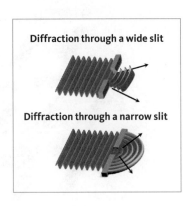

ISBN: 9780170368179

Interference and superposition

When two or more waves are travelling in the same region, they will **interfere** with each other and the **resultant wave** will be the **sum of the displacements** of the contributing waves. The process of adding waves together is called **superposition**.

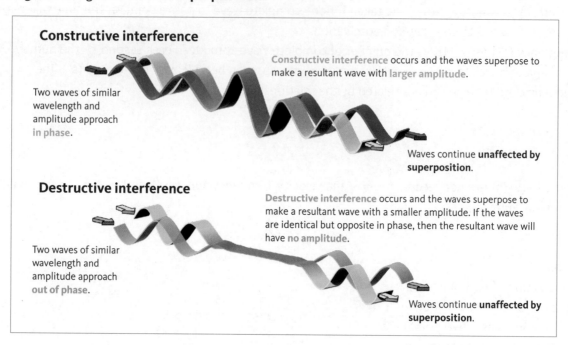

Constructive interference

Two waves of similar wavelength and amplitude approach in phase.

Constructive interference occurs and the waves superpose to make a resultant wave with larger amplitude.

Waves continue **unaffected by superposition**.

Destructive interference

Two waves of similar wavelength and amplitude approach out of phase.

Destructive interference occurs and the waves superpose to make a resultant wave with a smaller amplitude. If the waves are identical but opposite in phase, then the resultant wave will have no amplitude.

Waves continue **unaffected by superposition**.

Standing waves

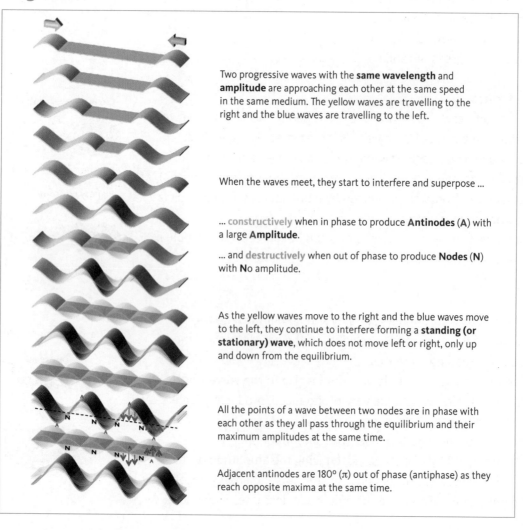

Two progressive waves with the **same wavelength** and **amplitude** are approaching each other at the same speed in the same medium. The yellow waves are travelling to the right and the blue waves are travelling to the left.

When the waves meet, they start to interfere and superpose ...

... constructively when in phase to produce **Antinodes** (**A**) with a large **Amplitude**.

... and destructively when out of phase to produce **Nodes** (**N**) with **N**o amplitude.

As the yellow waves move to the right and the blue waves move to the left, they continue to interfere forming a **standing (or stationary) wave**, which does not move left or right, only up and down from the equilibrium.

All the points of a wave between two nodes are in phase with each other as they all pass through the equilibrium and their maximum amplitudes at the same time.

Adjacent antinodes are 180° (π) out of phase (antiphase) as they reach opposite maxima at the same time.

 ISBN: 9780170368179

Standing waves can be produced when a wave reflects and the outgoing wave interferes with the reflected wave, for example mechanical waves on guitar strings, water waves in a bowl, sound waves inside a musical instrument such as the hollow pipe of a flute, or electromagnetic waves inside a microwave.

Extension concepts for scholarship: Oil on water — thin film interference

When white light is incident on a thin film of oil resting on water:

- some of the light is reflected from the surface of the oil and undergoes a **180° (π) phase change**, as the oil has a higher refractive index than the air, and
- some of the light is refracted into the oil and is not phase changed.

The refracted light is then incident on the oil-water boundary, where:

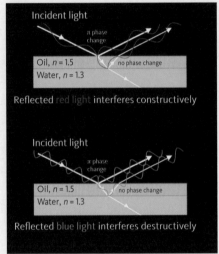

- some of the light is once again reflected, but it is **not phase changed**, as water has a lower refractive index than oil, and
- some of the light is refracted into the water and lost.

The reflected ray travels back towards the surface of the oil, where it is refracted and recombines with the light originally reflected from the surface of the oil.

- When the path difference results in the two waves being in phase, they constructively interfere and the colour of that wavelength is seen, for example the red wave in the diagram.
- When the path difference results in the two waves being out of phase, they destructively interfere and the colour of that wavelength disappears, for example the blue wave in the diagram.

The path difference varies with the angle of incidence of the light and the thickness of the film, which affects what colour is seen at different parts across the film.

Exercise 3A

1 Describe the similarities and differences between mechanical and electromagnetic waves and give examples of each type. (A)

_____

_____

_____

_____

_____

2 The diagram shows how the air particles move as a travelling sound wave moves through them and represents the particles motion using a sine wave.

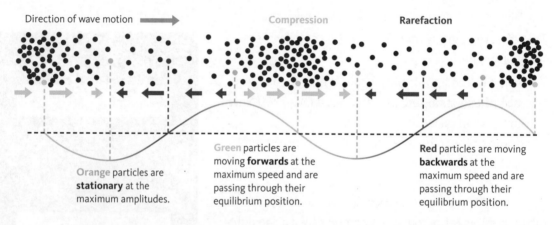

Direction of wave motion Compression Rarefaction

Orange particles are **stationary** at the maximum amplitudes.

Green particles are moving **forwards** at the maximum speed and are passing through their equilibrium position.

Red particles are moving **backwards** at the maximum speed and are passing through their equilibrium position.

a Compare and contrast longitudinal and transverse wave motion. (M)

_____

_____

_____

_____

_____

_____

_____

b Describe how the position of the air particles and the sine wave would look if the amplitude of the oscillation increased. (A)

_____

_____

_____

_____

_____

_____

c Describe how the position of the air particles and the sine wave would look if the frequency increased. (A)

_____

_____

_____

_____

_____

3 Diffraction effects become more significant as the gap through which the waves pass decreases.

a State what stays the same regardless of the size of the gap. (A)

_____

_____

_____

b Explain what happens to the amplitude of the diffracted waves as the size of the gap decreases. (M)

_____

_____

_____

_____

4 Explain the conditions necessary for a standing wave to form. (M)

_____

_____

_____

_____

_____

_____

_____

5 CD, DVD and Blu-ray disks store data as little pits in the surface of the disk. The pits are 1.95×10^{-7} m deep, which is exactly ¼ of the wavelength of the incident light.

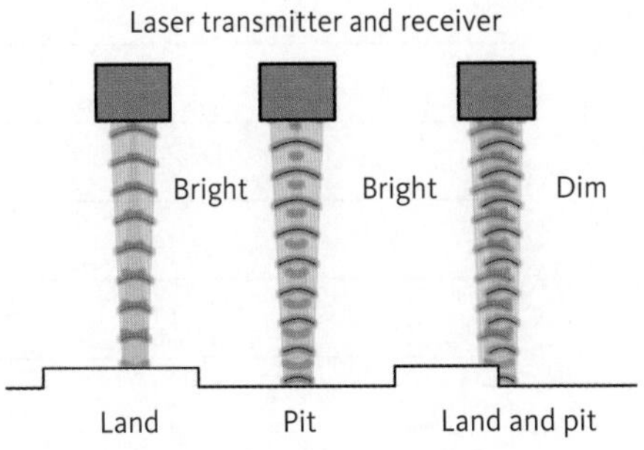

Laser transmitter and receiver

Bright   Bright   Dim

Land   Pit   Land and pit

a Show that the laser has a wavelength of 780 nm and state the colour. (A)

_____

_____

_____

_____

When the laser shines into a pit or onto a land (the surface of the disk), a bright reflection is detected by the sensor. But when the laser passes over an edge, the detected brightness decreases significantly.

b By considering the path difference between the light reflected from a pit and light reflected from a land, explain why there is a decrease in brightness when the laser is incident upon an edge. (E)

_____

_____

_____

_____

_____

_____

_____

_____

_____

The width of the laser beam when it hits the surface of the disk depends on the diameter of the aperture (hole) through which the laser is emitted, as shown in the diagrams below.

Wide aperture   Narrow aperture   Very narrow aperture

c Explain why the aperture of the laser must be significantly larger than the wavelength of the light to produce a narrow beam. (M)

_____

_____

_____

_____

_____

_____

_____

_____

A Blu-ray player uses a violet laser of frequency 7.407×10^{14} Hz and the disks can store significantly more data as a result.

d Discuss how changing the colour increases the amount of data that can be stored on a disk. In your answer you should consider the following:

- the effect of the wavelength on the diameter of the beam
- the distance between the tracks
- the size and depth of the pits.

_____

_____

_____

_____

_____

_____

_____

_____

_____

_____

_____

_____

Scholarship question

6 Josh wears glasses to help him with his computer work, however the light from the screen reflects off the surface of the glasses preventing his colleagues from seeing his eyes. His optician suggests that he could have an anti-reflective coating put on the surface of the lenses.

By considering the reflections at the coating surface and the coating-glass interface, determine the thickness of the coating layer to prevent reflections occurring. Fully justify your answer and explain any limitations of the layer.

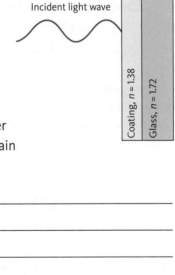

Incident light wave

Coating, $n = 1.38$

Glass, $n = 1.72$

_____

_____

_____

_____

_____

_____

_____

_____

3.1 Interference and diffraction

Double slit interference

When coherent (same phase, amplitude and frequency) waves pass through two adjacent narrow slits, diffraction causes the semi-circular wave fronts from each slit to overlap producing an interference pattern made up of a number of **bright (antinode)** and **dark (node)** fringes depending on the *path difference, pd*, between the waves travelling from the two slits.

For the *fringe separation, x*, to be as large as possible:

- the slit width must be similar to the size of the *wavelength*, λ, to cause significant diffraction
- the *separation of the slits, d*, should be small
- the *screen distance, L*, should be large.

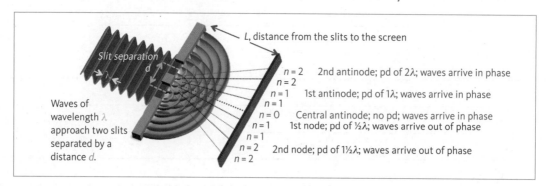

Antinodes

Antinodes form when there is a **whole number** of wavelengths path difference ($pd = 0\lambda, 1\lambda, 2\lambda, ...$) between the waves from the two slits, resulting in:

- the waves arriving in **phase**
- **constructive** interference, and
- a **large amplitude** oscillation (bright/light, loud/sound etc).

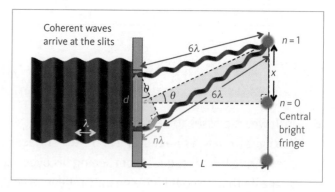

The diagram shows that when there is 1λ path difference, the waves form the $n = 1$ antinode (first bright fringe either side of the central fringe). So the nth bright fringe occurs when there is a path difference of $n\lambda$. The angle between the central bright fringe and the nth fringe (blue triangle) is given by:

$$\tan \theta = \frac{x}{L}$$

The angle formed by the slit separation and the path difference (green triangle) is given by:

$$\sin \theta = \frac{n\lambda}{d}$$

When $x \ll L$, the angle between the central bright fringe and the nth fringe is small ($< 10°$) so:

$$\sin \theta \approx \tan \theta$$

This **small angle approximation** gives the relationship for the distance to the nth antinode as:

$$n\lambda = \frac{dx}{L} \text{ where } n = 0, 1, 2, 3, ...$$

 ISBN: 9780170368179

which states:

$$n^{th} \text{ fringe x wavelength (m)} = \frac{\text{slit separation (m) x fringe distance from the centre (m)}}{\text{screen distance (m)}}$$

Nodes

Nodes form when there is a number of **half** wavelengths path difference (pd = ½λ, 1½λ, 2½λ, 3½λ, ...) between the wave paths from the two slits, resulting in:

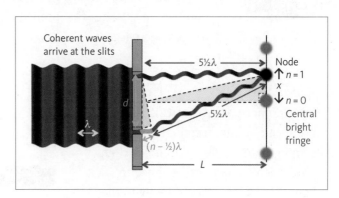

*   the waves arriving **out of phase**
*   **destructive** interference, and
*   **no amplitude** (dark/quiet/etc.),

which gives the relationship for the nth node as:

$$\left(n - \frac{1}{2}\right) \lambda = \frac{dx}{L} \text{ where } n = 1, 2, 3, ...$$

Notes:

* n is an integer, so must always be rounded to the nearest value.
*   When x is not significantly smaller than L so that θ exceeds the small angle approximation, the above formulae are not valid and must be solved in two steps using $\tan \theta = \frac{x}{L}$ and $\sin \theta = \frac{n\lambda}{d}$.
*   If the waves leaving the two slits are 180° (π radians) out of phase, the nodes and antinodes swap position.

Multiple slit interference and subsidiary maxima

Width of the beam

When a laser beam diffracts through a slit, it produces a wide low-intensity band. The width of the band is determined by the width of the slit.

Brightness increases

When a laser beam diffracts through two slits, the width of the pattern is determined by the width of the slits, so for the same slit width, the bright fringes fit into the same 'envelope' as the single-slit pattern.

As there are now twice as many slits for light to travel through, the fringes become brighter (greater intensity).

Fringes become more defined

As the number of slits increases, the bright fringes become narrower and more defined because a small change in angle means there is a greater probability that the waves from one of the slits will arrive π out of phase with the waves from another slit further along, resulting in some destructive interference. This causes the intensity to rapidly decrease.

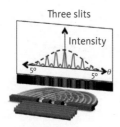

Three slits

Subsidiary maxima

Small subsidiary fringes begin to appear with multiple slits when light from all but one of the slits cancel out. For *N* slits there will be *N–2* subsidiary fringes.

The intensity of the fringes increases with increasing numbers of slits, but the intensity of the subsidiary fringes decreases.

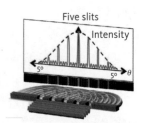

Five slits

Extension concepts for scholarship — Lloyd's mirror

Interference patterns can also be produced by light from a single slit shining onto a reflective surface then onto a screen and interfering with light that has travelled directly to the screen.

When a light wave is incident on the glass surface, it undergoes fixed end reflection and is phase changed by 180° (π radians).

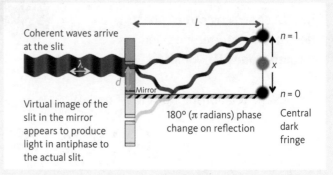

- When the direct wave and the reflected wave meet with a path difference of *nλ*, the phase change of the reflected wave means that they arrive out of phase resulting in destructive interference and a dark fringe appears where a bright fringe would normally be expected.
- When the direct wave and the reflected wave meet with a path difference of $(n - \frac{1}{2})\lambda$, the phase change of the reflected wave means that they arrive in phase resulting in constructive interference and a bright fringe appears where a dark fringe would normally be expected.

Even though all the waves are coming from the single slit, the reflection in the glass causes the reflected ray to appear to have come from the virtual image of the slit. So the experiment behaves as though there are two slits a distance 2*d* apart, producing waves in opposite phase. Hence the relationship between the wavelength, *λ*, and the distance between the central dark fringe and the *n*th dark fringes, *x*, is given by the formula (assuming a small angle approximation):

$$n\lambda = \frac{2dx}{L}$$

Worked example: Double slit interference

Vincent and Anshu carried out experiments to measure the speed of sound by creating an interference pattern using two loudspeakers placed 50.0 cm apart. They played a note of frequency 1.988 kHz and then measured the distance between the loud regions along a line 12.0 m from the speakers. They found that the first loud region was 4.37 m away from the central loud region.

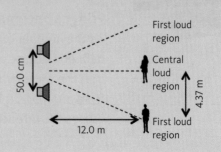

ISBN: 9780170368179

Vincent carried out the following calculation to determine the answer: $n\lambda = \dfrac{dx}{L}$ and $v = f\lambda$

hence $v = f\dfrac{dx}{nL} = 1988 \times \dfrac{0.500 \times 4.37}{1 \times 12.0} = 362$ m s^{-1}, which Vincent thought was too fast.

Anshu carried out an alternative solution and gained the answer 340 m s^{-1}, which is much closer to the quoted value for the speed of sound at room temperature.

Discuss why Vincent and Anshu get different answers. In your discussion you should also demonstrate how Anshu gained her answer.

Given	$d = 0.500$ m, $f = 1988$ Hz, $L = 12.0$ m, $n = 1$, $x = 4.37$ m
Unknown	$v = ?$
Equations	$n\lambda = d\sin\theta$, $v = f\lambda$, $\tan\theta = \dfrac{x}{L}$
Substitute	The double-slit formula $n\lambda = \dfrac{dx}{L}$ relies on the angle between adjacent antinodes to be small, less than about 10°. The formula is valid when $x \ll L$, however in this situation the distance between adjacent antinodes (x) is about ⅓ of the distance to the screen (L). As $\theta = \tan^{-1}\dfrac{4.37}{12.0} = 20.0°$, which exceeds the small angle approximation used to create the double-slit formula. The double slit formula is invalid. Using the diffraction angle formula, $n\lambda = d\sin\theta$, instead and combining it with the wave equation $v = f\lambda$ gives $v = \dfrac{fd\sin\theta}{n}$ Substituting in all the values gives:
Solve	$v = \dfrac{1988 \times 0.500 \times \sin 20.0}{1} = 340$ m s^{-1} (3 sf)

Exercise 3B

Take the speed of light as $c = 3.00 \times 10^8$ m s^{-1}.

1   Kieran produces an interference pattern on a screen 2.50 m from a double slit using a red laser with a wavelength of 6.60×10^{-7} m, which acts as a coherent source of light. The fifth bright fringe is 33.0 mm away from the central bright fringe.

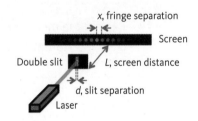
x, fringe separation
Screen
Double slit
L, screen distance
d, slit separation
Laser

   a   Show that the slits are separated by a distance of 0.250 mm. (A)

   _____

   _____

   b   Determine the maximum fringe number that would appear on a screen 80 mm wide and hence state how many bright fringes will appear on the screen. (M)

   _____

   _____

   _____

Using a yellow laser with the same slits, Kieran finds that the fifth bright yellow fringe is only 29.0 mm from the central bright fringe.

c Show that the wavelength of the yellow laser light is 580 nm. (A)

_____

_____

d Determine the distance at which the screen would need to be placed to make the second bright yellow fringe appear 23.2 mm from the central bright fringe. (A)

_____

_____

e With the aid of diagrams, explain why an interference pattern will not occur if the width of the slits is too great.

Wide slits   wave fronts   Narrow slits   wave fronts

_____

_____

_____

_____

_____

_____

_____

_____

f Explain the meaning of the term coherent and state the other conditions necessary for a clear interference pattern to be produced. (M)

_____

_____

_____

_____

g Explain why the central bright spot is referred to as the $n = 0$ antinode. (A)

_____

_____

_____

_____

    ISBN: 9780170368179

h Describe and explain the effect on the diffraction pattern caused by changing the following features of the experiment.

i Changing from red laser light (660 nm) to green laser light (530 nm). (M)

_____

_____

_____

_____

_____

ii Increasing the separation of the slits, d, for the same slit width. (M)

_____

_____

_____

iii Increasing the width of the slits for same slit separation. (M)

_____

_____

_____

iv Decreasing the distance between the slits and the screen. (M)

_____

_____

_____

_____

2 Claudia shines a blue laser of frequency 7.16×10^{14} Hz onto a double slit with a separation of 0.200 mm resulting in a narrow diffraction pattern appearing on a screen placed 3.00 m away from the double slits.

x, fringe separation

$n = 7\ 6\ 5\ 4\ 3\ 2\ 1\ 0\ 1\ 2\ 3\ 4\ 5\ 6\ 7$ Screen

Double slit $L = 3.00$ m

$d = 0.200$ mm

Laser, $f = 7.16 \times 10^{14}$ Hz

a Show that the wavelength of the laser light is 419 nm. (A)

_____

_____

b Show that the distance between the central bright fringe and the fourth bright fringe on the right-hand side $n = 4$ is 25.1 mm. (A)

_____

_____

_____

c Using the diagrams below, explain why the distance between the second bright fringe $n = 2$ on each side is also 25.1 mm. (M)

Kieran says that there is no light incident on the screen between the bright fringes, but Claudia argues that about the same amount of light is incident on the screen at a dark fringe as the bright fringe next to it.

d State who is correct and justify your answer. (M)

e Show that the distance between the central bright fringe and the sixth dark fringe $n = 6$ on the right hand side is 34.6 mm. (A)

f Using the diagrams below, explain why the distance between the third dark fringe $n = 3$ on each side is not 34.6 mm. Support your answer with calculations. (M)

 ISBN: 9780170368179

3 Michael shines a green laser onto a multiple slit and produces the following diffraction pattern on a screen 5.42 m from the slits. Each slit is separated by a distance of 0.250 mm. The pattern is made up of bright well-defined fringes (maxima) 10.0 mm apart with two smaller subsidiary maxima in between them.

● ● ˙ ● ˙ ● ˙ ˙ ● ˙ ˙ ● ˙ ˙ ● ˙ ˙ ● ˙ ˙ ● ˙ ● ˙ ● ˙ ˙ ●

a How many slits are required to create the pattern shown above? (A)

_____

b Determine the frequency of the light used to create the pattern. (M)

_____

_____

_____

_____

c Discuss what will happen to the brightness and definition of maxima and number of subsidiary maxima if Michael covers one of the slits. (E)

_____

_____

_____

_____

_____

_____

_____

_____

4 Lydia places two speakers 1.20 m apart and plays a coherent tone of frequency 1700 Hz through both speakers. When she walks across the room 6.0 m in front of the speakers, she passes through regions of loud sound and regions of quiet spread across a large angle. Lydia stops when she enters the third loud region, 3.46 m from the central loud region.

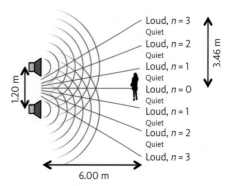

a Show that the speed of sound in the room is 340 m s⁻¹. (M)

_____

_____

_____

_____

_____

Lydia's friend Piper removes the connectors to one of the speakers, and reconnects the wires the wrong way around causing the speaker to operate 180° (π radians) out of phase with the first speaker.

b Discuss how this will affect the sound Lydia hears, 3.46 m from the centre. (E)

_____

_____

_____

_____

_____

_____

_____

_____

_____

_____

5 Chelsea has net curtains on her window and when looking through them one night she notices that the sodium street-light outside her house produces an interference pattern when viewed through the fabric as shown in the diagram. Chelsea realises that the fabric acts as a multiple slit with many lines. Microscope images of four fabric samples are shown alongside.

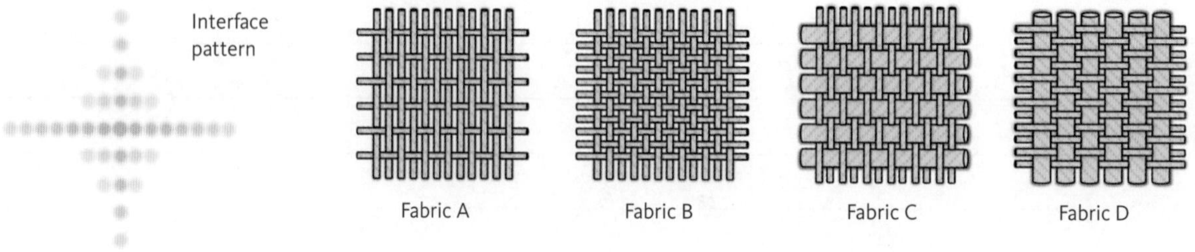

Interface pattern

Fabric A   Fabric B   Fabric C   Fabric D

a State which fabric sample most closely matches the fabric of her net curtains, and explain your answer. (E)

_____

_____

_____

_____

_____

_____

_____

_____

_____

_____

_____

_____

 ISBN: 9780170368179

b Determine the number of vertical threads per metre given that the sodium light has a frequency of 5.09 x 10¹⁴ Hz and the $n = 4$ antinodes on either side of the central fringe are 8.92 mm apart at a distance of 1.20 m from the curtains. (E)

_____

_____

_____

_____

_____

_____

Scholarship question

6 Marine animals such as whales and manatees use sound to help them navigate through the seas and oceans. Ships emit a low-frequency propeller sound, which should warn the animals of the danger, however marine animals near the surface are frequently hit by ships.

The diagram below shows sound waves travelling towards three whales both directly and after having been reflected off the surface of the water-air boundary.

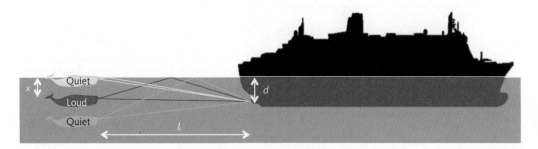

As a result of the two wave paths, the yellow and green whales cannot hear the propeller of the ship, whereas the red whale hears the noise of the propeller very clearly. Sound travels at 1500 m s⁻¹ in water but only 330 m s⁻¹ in air, so they experience a 180° (π radians) phase change when reflecting from the water-air boundary from inside the water.

a On the diagram, show the virtual source where the reflected sound waves appear to come from.

b Compare and contrast this situation with Young's double-slit experiment.

_____

_____

_____

_____

_____

_____

_____

_____

_____

c Explain why the yellow whale near the surface cannot hear the sound of the propellers.

_____

_____

_____

_____

_____

_____

_____

d Show that the distance, x, between the surface and a loud region is given by the formula:

$$x = \frac{(2n-1)vL}{4df}$$

_____

_____

_____

_____

_____

_____

_____

_____

_____

_____

_____

_____

The propeller of the ship is 8.00 m below the surface of the water and emits a sound of 1.245 kHz.

e Calculate the path difference between the sound waves if a whale, 44.0 m away from a ship and swimming at a depth of 5.0 m, is able to hear the sound of the propeller.

_____

_____

_____

_____

_____

_____

_____

 ISBN: 9780170368179

Diffraction gratings

Diffraction gratings are made by ruling narrow parallel lines of equal width onto glass or plastic (transmission gratings) or polished metal (reflection gratings). The ruled lines absorb the light so that only the light that strikes the surface is transmitted (or reflected) and produces a diffraction pattern. When white light shines on a diffraction grating, it produces a rainbow of colour, because different wavelengths are diffracted though different angles, for example CDs and DVDs act like a reflection grating due to the recording track lines on the discs.

Number of ruled lines, N, and line separation, d

There are typically more than 600 ruled lines per millimetre (600 000 lines per metre) on a diffraction grating. The distance, d, between each line is given by the relationship:

$$d = \frac{1}{N}$$

which can be written as:

$$\text{distance between the lines (m)} = \frac{1}{\text{number of lines per metre (m}^{-1})}$$

Diffraction grating formula

Due to the slits being extremely narrow, the bright fringes produced by a diffraction grating are very far apart. This means that the small angle approximation cannot be used, so the angle θ of the bright fringes to the $n = 0$ path must be determined using the formula:

$$n\lambda = d\sin\theta \text{ where } n = 0, 1, 2, 3, \dots$$

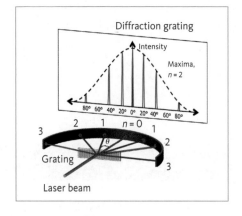

Note: This formula is true for all instances of diffraction — single slit, multiple slit and gratings. The bright fringes are very defined due to the large number of slits contributing to the interference pattern (see the notes on page 55 multiple-slit interference), and completely destructive interference between maxima means that there are no subsidiary maxima.

Number of bright fringes observed

The maximum possible angle of diffraction is 90° (along the face of the grating) and as sin 90° = 1, so $n\lambda \le d$, which can be rearranged to give the maximum number of bright fringes that can be produced by a grating:

$$n_{max} \le \frac{d}{\lambda}$$

Notes:
- As a bright fringe is a defined point, n_{max} must be **rounded down** to the nearest integer.
- The total number of bright fringes observed will be $2n + 1$, as there are fringes either side of the central bright fringe.

Worked example: Security vault

A movie director wants the special effects department to create a vault protected by laser beams. The vault is a square room 10.0 m on each side and the laser is set into the centre of one wall. The laser has a wavelength of 630 nm and shines through a diffraction grating so that 11 bright fringes just fit on the opposite wall.

Determine how many lines per mm must be drawn on the diffraction grating to achieve this effect. (E)

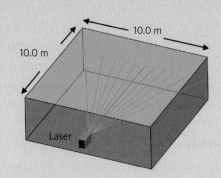

Given	$L = 10.0$ m, $\lambda = 630 \times 10^{-9}$ m, 11 bright fringes means there are $n = 5$ bright fringes either side of the central bright fringe, and $x = 5.00$ m from the centre, which means that we cannot use the small angle approximation.
Unknown	$N = ?$
Equations	$n\lambda = d\sin\theta$, $\tan\theta = \dfrac{x}{L}$ and $d = \dfrac{1}{N}$
Substitute	$\theta = \tan^{-1}\left(\dfrac{5}{10.0}\right) = 26.57°$
	And as $n\lambda = d\sin\theta$ then $d = \dfrac{n\lambda}{\sin\theta}$ so $N = \dfrac{\sin\theta}{n\lambda}$ and substituting in all the values gives:
Solve	$N = \dfrac{\sin 26.57°}{5 \times 630 \times 10^{-9}} = 141\,973$ m^{-1} = 142 lines per mm (3 sf)

Exercise 3C

1 Light from a red laser passes through a diffraction grating with a slit separation of 2.00×10^{-6} m and is diffracted through an angle of 38.68° to produce the second bright fringe ($n = 2$) from the centre.

a Calculate the wavelength of the light. (A)

_____

_____

The pattern is observed by projecting it onto a wall 6.8 m away from the grating,

b Calculate the distance of the bright fringe from the central bright fringe. (A)

_____

_____

2 A nitrogen lamp produces a bright turquoise (blue-green) line of wavelength 447.15 nm when viewed through a diffraction grating with exactly 333 lines per mm.

a How many lines are there per metre? (A)

_____

 ISBN: 9780170368179

b Show that the distance between each line in the grating is 3.00×10^{-6} m. (A)

_____

c Calculate the angle through which the fourth bright fringe ($n = 4$) is diffracted. (A)

_____

_____

d Determine the maximum order of fringe visible. (M)

_____

_____

3 Melissa buys some novelty diffraction glasses for a concert that make a 'rainbow' or spectrum of colours appear around any white light source that she looks at.

a Explain why the glasses have this effect and why blue light is diffracted the least. (M)

_____

_____

_____

_____

_____

_____

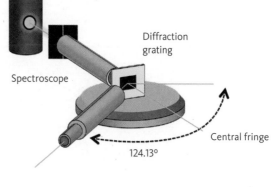

Each order of the spectrum is quite spread out and Melissa notices that red light in the second order spectrum coincides with violet light from the third order spectrum.

b Determine the wavelength of the violet light, given that the red light has a wavelength of 6.30×10^{-7} m. (A)

_____

_____

_____

4 Melissa uses a diffraction grating with 3.00×10^2 lines per mm to view light of frequency 5.093×10^{14} Hz through a spectroscope. She gradually rotates the spectroscope, revealing several bright orange fringes on either side of the central bright fringe. The last bright orange fringes on either side of the pattern are 124.13° apart.

a Show that the angle of the last bright orange fringes from the central bright fringe is 62.065°.

_____

_____

b Determine the order, *n*, of the last bright orange fringe. (M)

_____

_____

_____

_____

c Discuss the conditions necessary to form this particular order of bright fringe and explain why it is so narrow and well defined. (E)

_____

_____

_____

_____

_____

_____

_____

_____

_____

_____

d State how many bright fringes appear in the pattern. (A)

_____

_____

e Discuss how the number of bright fringes observed would change if a diffraction grating with more lines per mm were used. Hence determine the maximum number of lines per mm for a grating that would still produce the same number of fringes. (E)

_____

_____

_____

_____

_____

_____

_____

_____

_____

 ISBN: 9780170368179

5 Carolle is studying the light from the aurora australis. She uses a telescope to focus the light through a spectroscope resulting in the pattern shown below the photograph.

The lines in the pattern are emitted by excited oxygen molecules in the upper atmosphere, and when viewed through the spectroscope, a green line is observed at an angle of 33.90° away from the central bright fringe. A red line of the same order is observed at an angle of 40.20° from the central bright fringe. The two lines have a difference in wavelength of 87.8 nm.

a Determine the wavelength of the two lines. (E)

_____

_____

_____

_____

_____

_____

_____

_____

A green line of the same wavelength but in the next order is observed at an angle of 56.78°.

b Determine the orders of the two green lines. (E)

_____

_____

_____

_____

_____

_____

6 The information on a CD is 'read' using a red laser of frequency 3.846×10^{14} Hz, which shines onto the surface of the CD, normal to the surface. The CD diffracts the laser light resulting in a diffraction pattern on the case of the player, as shown in the diagram opposite. The case is 10.0 cm from the disc and the $n = 1$ antinodes are 11.96 cm apart.

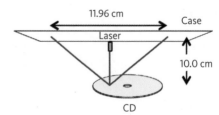

a Show that there are 658 tracks per mm on a CD. (E)

_____

_____

_____

_____

_____

There are 1389 tracks per mm on a DVD, far more than the tracks on the CD, allowing it to store more data, but this requires a higher frequency laser to be used. The first-order bright fringes are 38.82 cm apart.

b Show that the frequency of the laser light used in a DVD is 4.69×10^{14} Hz. (E)

_____

_____

_____

_____

_____

_____

_____

c Discuss what happens to the diffraction pattern if a CD is played in a DVD player. (E)

_____

_____

_____

_____

_____

_____

_____

_____

_____

_____

_____

 ISBN: 9780170368179

3.2 Acoustics

Sound waves

Sound is a longitudinal wave that travels through mechanical media as a series of compressions (high pressure) and rarefactions (low pressure). Sound waves will undergo reflection when they encounter a boundary with a higher or lower density. Consider a compression that travels down a pipe when you blow across the top of it.

Closed pipes — fixed or closed end reflection

When the compression pulse pushes on the solid end, the tube pushes back causing the compression pulse to travel back up the tube.

The air particles in the *reflected pulse*, are being pushed in the opposite direction to the air particles in the *incident pulse* (**180° phase change**) so the motions interfere destructively resulting in stationary air (**no amplitude**) at the closed end.

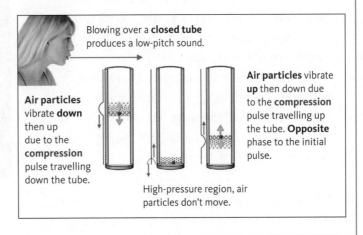

Blowing over a **closed tube** produces a low-pitch sound.

Air particles vibrate **down** then up due to the **compression** pulse travelling down the tube.

High-pressure region, air particles don't move.

Air particles vibrate **up** then down due to the **compression** pulse travelling up the tube. **Opposite** phase to the initial pulse.

Open pipes — free or open end reflection

The compression pulse rushes outwards at the open end leaving a void (low-pressure region) behind it which draws air down from the tube above.

The air rushing down the tube to fill the void is moving in the same direction as the escaping air (**same phase**) so interferes constructively (**large amplitude**) at the open end. The downwards motion of the air particles to fill the void causes the rarefaction to move upwards.

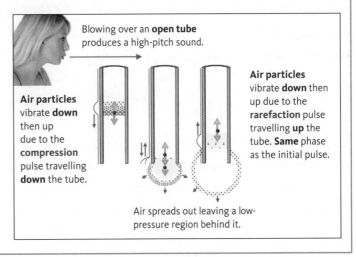

Blowing over an **open tube** produces a high-pitch sound.

Air particles vibrate **down** then up due to the **compression** pulse travelling **down** the tube.

Air spreads out leaving a low-pressure region behind it.

Air particles vibrate **down** then up due to the **rarefaction** pulse travelling **up** the tube. **Same** phase as the initial pulse.

Resonance

All objects will oscillate when a driving force is applied to them. For example:

- Pluck a guitar string and a wave travels down the string and reflects from the end. The note produced depends on: the length, the thickness and the tension of the string.
- Blowing across the mouthpiece of a flute causes the air to vibrate at the end sending a longitudinal wave down the pipe, which reflects from the end of the pipe. The note produced depends on: the gas in the pipe, the length of the pipe, and the temperature of the gas.

Where the incident and reflected waves meet, they superpose resulting in constructive and destructive interference.

- If the frequency of vibration of the driving force matches the natural frequency of an object, then interference produces a standing wave with a large amplitude, which can be heard and the object resonates. The natural or resonant frequencies of an object depend on the number of waves that 'fit' in the length of the object (see the table on the next page). Play the resonant frequency of a wine glass loud enough and it will shatter due to the amplitude of the vibrations!

- If the frequency of the vibration of the driving force does not match the natural frequency, then the wavelength of the wave does not 'fit' the object and the energy of the vibration is quickly lost so little or no sound is heard.

Acoustic terminology

Frequency (f) — a quantitative measure of the number of vibrations each second, e.g. 261.6 Hz.

Pitch — a qualitative measure of the frequency, e.g. middle C.

Fundamental (f_1) — the longest wave that 'fits' on a string or pipe and produces the lowest note with the greatest energy. It is also known as the 1st harmonic.

Harmonics (f_n) — vibrations having a frequency that is a multiple of the fundamental frequency. The frequency of the nth harmonic is given by the formula:

$$f_n = n \times f_1 \quad \text{where } n = 1, 2, 3, 4 \ldots$$

Overtones — when resonance occurs and several harmonics are produced at the same time as the fundamental. Closed pipe instruments cannot produce even-numbered harmonics, so have fewer overtones and so sound 'thinner' or 'less rich'.

Timbre — the character or quality of a musical instrument due to the mixture and relative intensity of the overtones. Timbre depends on a number of factors, such as the shape of an instrument and the materials from which the instrument is made.

Loudness — a subjective quantity that depends upon the intensity of the sound and the sensitivity of the listening device.

 ISBN: 9780170368179

Strings and pipes

Properties	Strings (and pipes closed at both ends)	Open pipes	Closed pipes
Description	**Nodes** form at each end because **fixed end reflection** results in **180° phase change** causing destructive interference between the incident and reflected waves.	**Antinodes** form at each open end because **free end reflection** results in **no phase change** causing constructive interference between the incident and reflected waves.	A **node** forms at the closed end because of fixed end reflection and an **antinode** forms at the open end because of free end reflection.
1st harmonic	$\lambda = 2L$ and $f = \frac{v}{\lambda}$ so $f_1 = \frac{v}{2L}$ **Fundamental**	$\lambda = 2L$ and $f = \frac{v}{\lambda}$ so $f_1 = \frac{v}{2L}$ **Fundamental**	$\lambda = 4L$ and $f = \frac{v}{\lambda}$ so $f_1 = \frac{v}{4L}$ **Fundamental**
2nd harmonic	$\lambda = \frac{2L}{2}$ and $f = \frac{v}{\lambda}$ so $f_2 = \frac{2v}{2L}$ *1st overtone*	$\lambda = \frac{2L}{2}$ and $f = \frac{v}{\lambda}$ so $f_2 = \frac{2v}{2L}$ *1st overtone*	2nd harmonic not produced as it must be half the wavelength of the 1st harmonic which would result in a node forming at the open end.
3rd harmonic	$\lambda = \frac{2L}{3}$ and $f = \frac{v}{\lambda}$ so $f_3 = \frac{3v}{2L}$ *2nd overtone*	$\lambda = \frac{2L}{3}$ and $f = \frac{v}{\lambda}$ so $f_3 = \frac{3v}{2L}$ *2nd overtone*	$\lambda = \frac{4L}{3}$ and $f = \frac{v}{\lambda}$ so $f_3 = \frac{3v}{4L}$ *1st overtone*
4th harmonic	$\lambda = \frac{2L}{4}$ and $f = \frac{v}{\lambda}$ so $f_4 = \frac{4v}{2L}$ *3rd overtone*	$\lambda = \frac{2L}{4}$ and $f = \frac{v}{\lambda}$ so $f_4 = \frac{4v}{2L}$ *3rd overtone*	4th harmonic not produced as it must be a quarter of the wavelength of the 1st harmonic which would result in a node forming at the open end.
5th harmonic	$\lambda = \frac{2L}{5}$ and $f = \frac{v}{\lambda}$ so $f_5 = \frac{5v}{2L}$ *4th overtone*	$\lambda = \frac{2L}{5}$ and $f = \frac{v}{\lambda}$ so $f_5 = \frac{5v}{2L}$ *4th overtone*	$\lambda = \frac{4L}{5}$ and $f = \frac{v}{\lambda}$ so $f_5 = \frac{5v}{4L}$ *2nd overtone*
n^{th} harmonic	$f_n = \frac{nv}{2L}$ where $n = 1, 2, 3, 4, 5, ...$	$f_n = \frac{nv}{2L}$ where $n = 1, 2, 3, 4, 5, ...$	$f_n = \frac{nv}{4L}$ where $n = 1, 3, 5, ...$
Speed of wave	The speed of the wave in the string, v, is related to the tension in the string, T, and the mass per unit length, μ: $v = \sqrt{\frac{T}{\mu}}$	Sound travels faster in: warmer gases, low-mass gases, e.g. helium. Humidity increases the speed of sound (by about 0.1–0.6%), as water molecules are lighter then oxygen and nitrogen molecules.	
Timbre	Rich sound due to the number of overtones produced.		Thinner sound due to some harmonics missing, so fewer overtones.

Worked example: The trombone

The trombone can be modelled on an open pipe. The trombonist can change the note by either changing the length or changing the pressure of the air to produce different harmonics. When the slide of the trombone is pushed all the way in, the effective length of the open pipe is 295 cm long and produces a note of frequency 175 Hz when the third harmonic is sounded.

a   Draw a diagram to show the third harmonic in the model open pipe.

b   Calculate the speed of sound in the trombone.

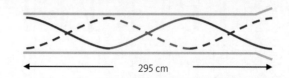

Given	$d = 2.95$ m, $f = 175$ Hz, $n = 3$
Unknown	$v = ?$
Equations	$v = f\lambda$ and from the diagram it can be seen that $L = \frac{3}{2}\lambda$
Substitute	$\lambda = \frac{2L}{3} = \frac{2 \times 2.95}{3} = 1.97$ m
Solve	$v = f\lambda = 175 \times 1.97 = 344$ m s^{-1} (3 sf)

Extension concepts for scholarship — Intensity and the inverse square law

Intensity is defined as the power per unit area and is important when considering how the energy associated with a wave spreads out as the wave travels away from a source.

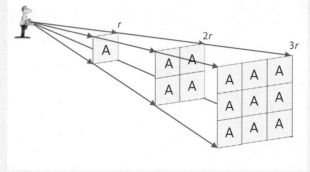

Consider a person clapping in the centre of a stadium. The sound will travel away from them in all directions as a spherical wave.

The energy of the sound is spread over the surface of the sphere, so as the distance increases, the energy per unit area decreases.

When the sound has travelled a distance r, the energy is spread over an area A. By the time the sound has travelled $3r$, the same energy is spread over an area of $9A$.

This shows that the intensity of a sound obeys the **inverse square law**, which states that 'The intensity is inversely proportional to the square of the distance from the source', and applies to many aspects of physics such as light, Newton's law of gravitation, electric fields and magnetic fields.

For sound, the intensity can be expressed mathematically as:

$$I = \frac{P}{4\pi r^2}$$

which states:

$$\text{intensity (W m}^{-2}) = \frac{\text{power (W)}}{\text{area of a sphere (m}^2)}$$

Exercise 3D

Speed of sound in air is $v = 340$ m s^{-1} at room temperature.

1   The diagrams below show sets of **fixed strings**, **open pipes** and **closed pipes**. Determine the harmonic shown in each of the following diagrams (hint: count the number of 'fundamental shapes') and state the wavelength in terms of the pipe length L. The first row has been done for you.

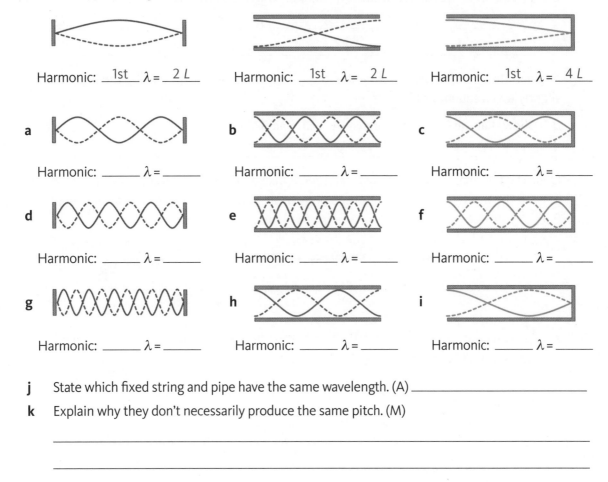

Harmonic: __1st__ $\lambda = $ __2 L__   Harmonic: __1st__ $\lambda = $ __2 L__   Harmonic: __1st__ $\lambda = $ __4 L__

a   Harmonic: _____ $\lambda = $ _____   b   Harmonic: _____ $\lambda = $ _____   c   Harmonic: _____ $\lambda = $ _____

d   Harmonic: _____ $\lambda = $ _____   e   Harmonic: _____ $\lambda = $ _____   f   Harmonic: _____ $\lambda = $ _____

g   Harmonic: _____ $\lambda = $ _____   h   Harmonic: _____ $\lambda = $ _____   i   Harmonic: _____ $\lambda = $ _____

j   State which fixed string and pipe have the same wavelength. (A) _____

k   Explain why they don't necessarily produce the same pitch. (M)

_____

_____

_____

_____

2   Josh is playing his solid silver flute in preparation for a concert. The blowhole on the flute acts as an open end, so the flute can be modelled on an open pipe. The player can change the pitch of the note by opening or closing the note holes along the length of the body and so change the effective length of the flute. When all the note holes are covered, the flute behaves like a 650 mm long **open pipe**.

a   On the diagram below, draw the fundamental wave inside the flute when all the note holes are covered. Label the nodes and antinodes. (A)

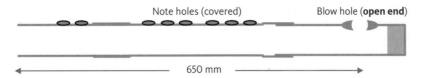

Note holes (covered)   Blow hole (**open end**)

650 mm

b Explain why the wavelength of the fundamental is 1.300 m. (A)

_____

_____

_____

_____

c Calculate the frequency of the fundamental when the flute is first played. (A)

_____

_____

As the flute is played, the instrument and the air inside it warm up, causing the note to change slightly to a frequency of 264 Hz.

d Show that the speed of sound in the warm air 343 m s^{-1}. (A)

_____

Once the flute has warmed up, Josh can tune it back to the original frequency of 261.5 Hz by moving the head joint into or out of the body of the flute. This changes the overall length of the flute.

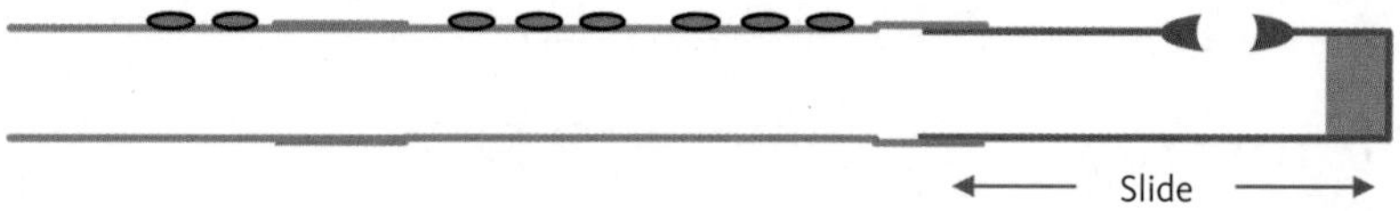

Body   Head joint

Slide

e Explain how Josh should move the head of the flute so that he can play the original frequency of 261.5 Hz. (M)

_____

_____

_____

_____

f On the diagram below, show what will happen to the shape of the fundamental wave if Josh opens the holes on the lower half of the flute.

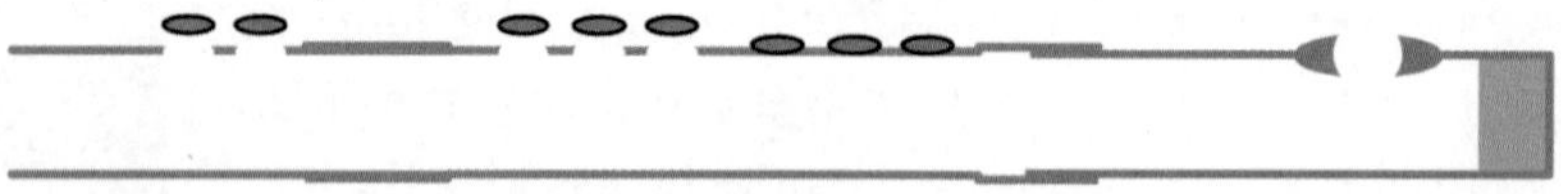

Open holes   Closed holes

g Explain how opening half the holes will affect the pitch of the note produced. (M)

_____

_____

_____

_____

 ISBN: 9780170368179

3   Georgia plays the clarinet, a reed instrument made from wood that can be modelled on a long pipe which is closed at the mouthpiece end and open at the bell.

By changing her embouchure (mouth shape) and how hard she blows, she can make the clarinet play different harmonics without changing her finger positions.

a   On the diagram below, draw the third harmonic wave inside the clarinet when all the note holes are covered. Label the nodes and antinodes.

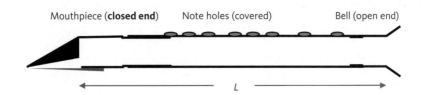

b   Determine the length, L, of the clarinet given that the third harmonic has a frequency of 494 Hz. (M)

_____

_____

_____

c   Calculate the frequency of the fundamental note, given that $f_n = nf_1$. (A)

_____

d   Explain why Georgia cannot produce the second harmonic on her clarinet. (M)

_____

_____

_____

_____

_____

_____

_____

e   Discuss how a standing wave forms inside the clarinet. (E)

_____

_____

_____

_____

_____

_____

Josh on his flute (see question **1**) and Georgia on her clarinet are tuning their instruments so they both play the same note while listening to a tuner.

f Discuss why the flute and clarinet don't sound the same even though they are both playing the same note. (E)

_____

_____

_____

_____

_____

_____

_____

_____

_____

_____

_____

_____

_____

_____

_____

4 In the mid-1800s, the famous classical guitar maker Antonio De Torres Jurado set the string length of the guitar as 650 mm. When any string is plucked, the main note that is heard is the fundamental, but the string will also produce much quieter harmonics at the same time. A skilled musician can make a string produce a loud harmonic instead without changing the length of the string.

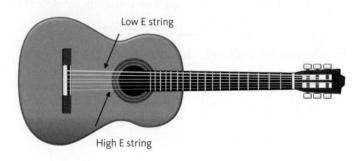

Low E string

High E string

a On the diagram below, draw the 4th harmonic on the high E string. (A)

 ISBN: 9780170368179

b What is the relationship between the wavelength of the 4th harmonic and the wavelength of the fundamental wave on the same string? (A)

_____

_____

c Show that the speed of the wave down the high E string is 428 m s⁻¹ when the 4th harmonic has a frequency of 1318.4 Hz. (M)

_____

_____

_____

d Calculate the tension in the E string if its mass per unit length is 389 x 10⁻⁶ kg m⁻¹. (A)

_____

_____

There are six strings on a classical guitar. The three high-pitch strings are made from fine nylon, whereas the three lower-pitch strings are made from nylon threads wrapped in wire. All the strings in a classical guitar are the same length and have the same tension. The fundamental frequency of the high E string is 329.6 Hz and the fundamental frequency of the low E string is 82.4 Hz.

e Discuss how the low E string is able to produce a frequency of 82.4 Hz even though it is the same length and tension as the high E string. (E)

_____

_____

_____

_____

_____

_____

_____

_____

_____

_____

In a search for greater volume, modern guitar builders have increased the length to 660 mm.

f Explain how increasing the length increases the volume of the guitar even though the same type of strings are used. (M)

_____

_____

_____

_____

_____

_____

Beats

Beats are produced when two waves of similar frequency and amplitude are heard together causing the listener to hear a gradual 'rise and fall' in loudness of the sound. In the example below, the blue wave has a slightly lower frequency (longer wavelength) than the red wave. As the waves arrive, they combine in and out of phase resulting in alternating periods of constructive and destructive interference causing the amplitude and loudness to vary.

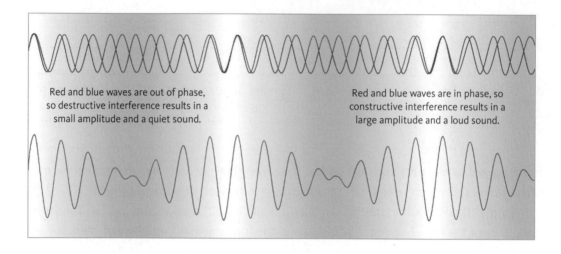

Red and blue waves are out of phase, so destructive interference results in a small amplitude and a quiet sound.

Red and blue waves are in phase, so constructive interference results in a large amplitude and a loud sound.

Beat frequency

The rate at which the loudness rises and falls is called the beat frequency and can be determined by finding the difference between the frequencies of the two notes.

$$f_{beats} = |f_1 - f_2|$$

(The vertical lines in the formula mean 'absolute value' and so the answer is always recorded as positive.)

Musicians use this physical phenomena to tune their instruments, because the rate of the beating decreases as the two frequencies approach the same value.

 ## Worked example: Tuning up!

Two musicians are tuning their instruments prior to a performance. The oboe player plays constant pitch (note A above middle C) with a frequency of 440.0 Hz. A clarinet playing at the same time is out of tune due to being slightly too high pitched and so the player hears beats with a frequency of 2.5 Hz. Describe and explain why beats occur and determine the frequency of the clarinet.

Solution

The cause of beats is a basic definition that must be committed to memory, but when answering a question the concepts need to be applied to the situation.

 ISBN: 9780170368179

When beats occur, the clarinet player will hear a gradual 'rise and fall' in loudness of the sound due to the sound waves from the oboe and the clarinet interfering. For beats to be produced, the frequency of the sound wave from the oboe and the clarinet must be similar and heard together. This slight difference in the frequency means that the waves alternate between being in and out of phase, resulting in alternating periods of constructive and destructive interference, which is why the amplitude and loudness vary.

Given	$f_0 = 440.0$ Hz, $f_B = 2.5$ Hz
Unknown	$f_C = ?$
Equations	$f_{beats} = |f_1 - f_2|$ and the question tells us that the clarinet is higher frequency so:
Substitute	$2.5 = f_C - 440.0$ so
Solve	$f_C = 442.5$ Hz

Exercise 3E

1   Eric plays in a rock band and uses a digital tuner to tune the lowest E string on his electric guitar. Once this is in tune, he then shortens the string by pressing on the fifth fret to produce a higher frequency of 110.0 Hz.

   a   Explain why shortening the E string produces a higher frequency. (A)

   _____

   _____

   _____

The A string on the guitar has a wavelength of 1.320 m and is untuned so produces a note of frequency 107.0 Hz.

   b   Calculate the speed of the wave in the A string if the fundamental frequency of the untuned string is 107.0 Hz. (A)

   _____

   _____

Eric uses the 110.0 Hz note from the E string to tune the A string on the guitar.

   c   Calculate the beat frequency that will be heard when the two strings are played at the same time. (A)

   _____

   _____

d Discuss how increasing the tension in the A string will affect the beat frequency. (E)

_____

_____

_____

_____

_____

_____

_____

Eric over-tightens the A string so that it produces a note higher than the 110.0 Hz reference note. When he tests the two strings again, he now hears a beat frequency of 1.2 Hz.

e Calculate the new frequency of the A string. (A)

_____

_____

2 The pan flute is a simple musical instrument made from a set of pipes that are closed at one end. Sound can be produced by blowing across the open end. When making a set of pipes, the craftsman tests the new pipes against a set of reference pipes to make sure they are in tune. The reference pipe is $L_R = 35.23$ cm long and produces a frequency $f_R = 243.4$ Hz. The new pipe is slightly longer and when it is played at the same time as the reference pipe, beats are heard with a frequency of $f_B = 3.2$ Hz. The speed of sound in air is $v = 343.0$ m s^{-1}.

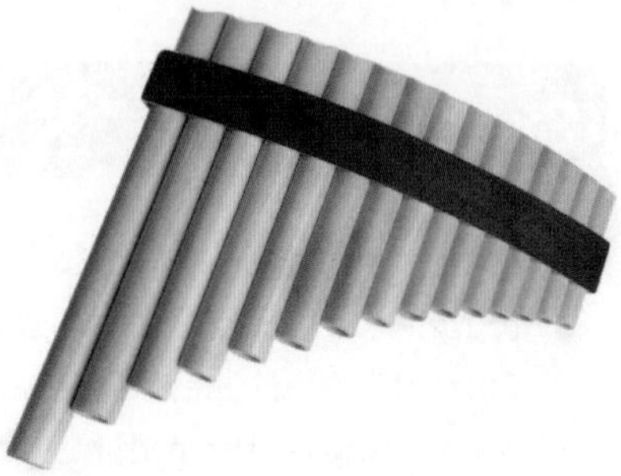

a Determine the difference in the length of the two pipes. (E)

_____

_____

_____

_____

_____

As the craftsman continues to blow into the new pipe, the pipe and air inside it warm up causing the speed of sound to increase.

b Explain how this will affect the beat frequency when he tests it against the reference pipe, which is still 'cold'. (M)

_____

_____

_____

_____

_____

 ISBN: 9780170368179

Once the pan flutes have been constructed and tuned, the craftsman notices that blowing across the longest pipe also causes some of the other pipes to make a sound at the same time.

c   Explain why a pipe ⅓ of the length produces sound at the same time, but a pipe ½ the length does not.

_____

_____

_____

_____

_____

_____

_____

_____

_____

_____

3   Josh is restringing his bass guitar in preparation for a jazz competition. He buys two different sets of strings of the same length to compare how 'heavy' and 'light' strings sound. The mass per unit length of the heavy string is 1.049 times greater than the mass per unit length of the light string. He attaches both the heavy and the light E strings to his bass guitar and tightens them up to the same tension.

a   Show that the wave travels 1.024 times faster in the light string. (M)

_____

_____

_____

_____

_____

When he plays the heavy and light strings at the same time, he hears beats with a frequency of 0.99 Hz.

b   Show that the frequency of the notes produced by the heavy and light strings are 40.93 Hz and 41.92 Hz respectively. (E)

_____

_____

_____

_____

_____

c Explain why beats occur. (E)

_____

_____

_____

_____

_____

_____

_____

Josh then tests each string with a tuning fork and hears beats, indicating that both strings are out of tune with the frequency of the tuning fork. The light string is higher than the correct frequency and the heavy string is lower than the correct frequency. The ratio of the beat frequencies of the two strings when played with the tuning fork is:

$$\frac{f_{bL}}{f_{bh}} = 2.67$$

d Using the beats formula $f_B = |f_1 - f_2|$ show that the tuning fork has a frequency of $f_T = 41.2$ Hz. (E)

_____

_____

_____

_____

_____

_____

_____

_____

  ISBN: 9780170368179

3.3 Doppler effect with mechanical waves

The Doppler effect is observed when an object producing waves travels towards or away from an observer resulting in a change in the observed frequency, for example the pitch of a siren falls as an ambulance drives passed.

Stationary wave source

Consider a penguin in the middle of a pool, acting as a **source of waves** by flapping its wings up and down with a frequency f. It produces circular wave fronts of wavelength λ, which travel away with a wave velocity v_{wave}, or v_w, as shown in the diagram.

The waves created by the penguin will reach all three observers with the same frequency as the penguin flaps.

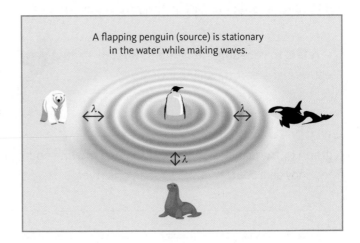

A flapping penguin (source) is stationary in the water while making waves.

Moving wave source

The penguin starts moving to the right with speed v_{source}, or v_s. The **speed of the waves**, v_w, **remains constant** and is unaffected by the speed of the source, v_s. So after one time period T, each wave will travel one wavelength, $\lambda = v_w T$.

However, when the next wave front is emitted, the source has moved a small distance $d = v_s T$ away from the polar bear and closer to the orca.

Consequently:

* The wavelength behind the penguin increases as:

$$\lambda_{behind} = v_w T + v_s T$$

hence the polar bear observes the waves arriving with a lower frequency than the receding source.

* The wavelength in front of the penguin decreases as:

$$\lambda_{in\ front} = v_w T - v_s T$$

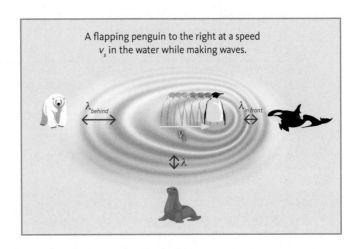

A flapping penguin to the right at a speed v_s in the water while making waves.

hence the orca observes the waves arriving with a greater frequency than the approaching source.

* The penguin is moving perpendicular to the line of sight of the seal, so the wavelength perpendicular to the direction of motion of the penguin remains the same and the seal observes the waves arriving with the same frequency as the source.

The relationship between the emitted frequency, f, and the observed frequency, f', is given by:

$$f' = f \frac{v_w}{v_w \pm v_s}$$

ISBN: 9780170368179

which states that:

$$\begin{matrix} \text{observed frequency} \\ \text{(Hz)} \end{matrix} = \begin{matrix} \text{emitted frequency} \\ \text{(Hz)} \end{matrix} \times \frac{\text{velocity of wave (m s}^{-1}\text{)}}{\text{velocity of wave (m s}^{-1}\text{)} \pm \text{velocity of source (m s}^{-1}\text{)}}$$

Remember ± using the dart analogy from field theory.

'**+**' refers to **receding** (remember: the tail feathers of a dart travelling away).

'**–**' refers to **approaching** (remember: the long shart tip of the dart).

Graphs of the Doppler effect

A graph of frequency against position of a moving source reveals the change in the observed frequency as a source moves past an observer.

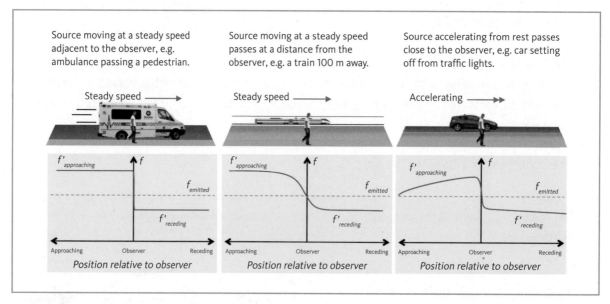

Note: From the graphs, it can be seen that the change in frequency is not 'symmetrical' above and below the emitted frequency. The approaching Doppler shift, Δf_a, is greater than the receding shift, Δf_r.

Worked example: Speeding

A car travelling west along a road sounds its horn with a frequency of 412.0 Hz, but an observer standing further along the road hears a frequency of 448.6 Hz as the car approaches. Speed of sound in air $v_w = 343$ m s^{-1}.

a   Use a diagram to explain why the wavelength of the sound decreases in front of the car.

To draw an accurate Doppler effect sketch:
1   Mark four dots each 0.2 cm apart.
2   Use drawing compasses to draw:
   a   4.0 cm circle centred on the dot 1
   b   3.0 cm circle centred on the dot 2
   c   2.0 cm circle centred on the dot 3
   d   1.0 cm circle centred on the dot 4
3   Draw an arrow from dot 1 to dot 4 to show the direction of motion and label it v_s.

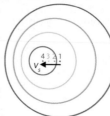

Waves in front of the car are close together as each successive wavefront is emitted closer than the last.

Waves behind the car are further apart as each successive wavefront is emitted from further away than the last.

b Determine the speed of the car.

Given	$f = 412.0$ Hz, $f' = 448.6$ Hz, $v_w = 343$ m s⁻¹
Unknown	$v_s = ?$
Equations	$f' = f\dfrac{v_w}{v_w \pm v_s}$ and as the car is approaching, the '–' term is used.

Substitute

Putting the numbers in first can make this rearrangement easier as follows:

$448.6 = 412.0 \left(\dfrac{343}{343 - v_s}\right)$ then multiply

412.0×343 to simplify the right-hand side

$448.6 = \dfrac{141316}{(343 - v_s)}$ rearranging becomes

$343 - v_s = \dfrac{141316}{448.6} = 315.016$ so

A neat algebraic solution can also be used to solve this problem:

Rearranging $f' = f\dfrac{v_w}{v_w - v_s}$ becomes

$(v_w - v_s) = \dfrac{f}{f'} v_w$ which becomes

$v_s = (v_w - \dfrac{f}{f'} v_w)$ which simplifies to become

$v_s = v_w\left(\dfrac{f' - f}{f'}\right)$ so

Solve

$v_s = 343 - 315.016 = 27.98$

$v_s = 28.0$ m s⁻¹ (3 sf)

$v_s = 343\left(\dfrac{448.6 - 412.0}{448.6}\right) = 27.98$

$v_s = 28.0$ m s⁻¹ (3 sf)

Exercise 3F

1 A terrapin (like a small turtle) is caught on a small branch in the water. Its flippers splash the water with a frequency of 2.0 Hz producing waves that travel away from the front of its body at 2.0 cm s⁻¹.

a Calculate the wavelength of the waves produced by the stationary terrapin. (A)

_____

It breaks free of the small branch and swims east at a steady speed of 1.2 cm s⁻¹. Its position is shown at times $t = 0.0, 0.5, 1.0, 1.5$ and 2.0 s in the diagram.

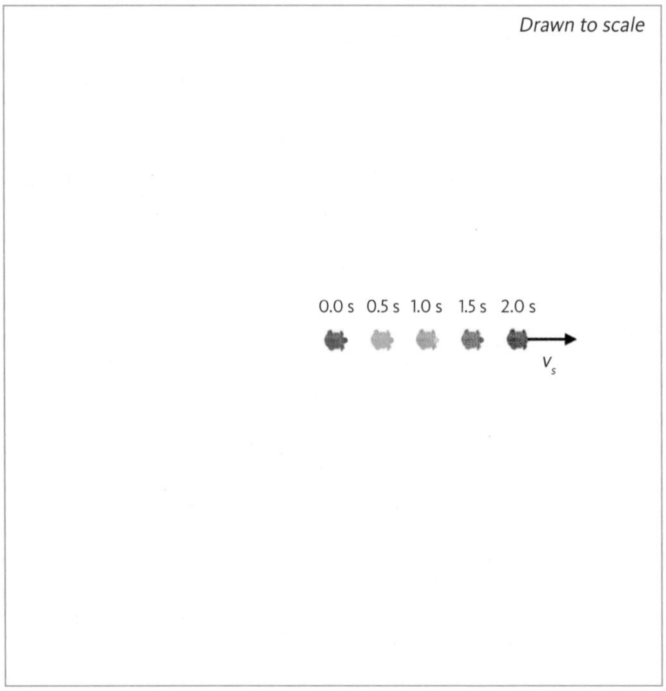

b Show that the wave emitted at $t = 0.0$ s (blue terrapin) will have travelled 4.0 cm by the time the terrapin reaches the $t = 2.0$ s position (black terrapin).

_____

_____

c Using a drawing compass, accurately draw in this circular wavefront around the blue terrapin.

Drawn to scale

0.0 s 0.5 s 1.0 s 1.5 s 2.0 s

v_s

d Using a drawing compass, accurately draw in the circular wavefronts to show where the waves emitted at $t = 0.5$ (green), 1.0 (orange) and 1.5 s (red) will be by the time the terrapin reaches $t = 2.0$ s (black). (M)

e Calculate the observed wavelength behind the terrapin using $\lambda_{behind} = v_w T + v_s T$ (check your answer by measurement). (A)

_____

_____

f Calculate the observed frequency behind the terrapin using $f' = f\dfrac{v_w}{v_w + v_s}$. (A)

_____

_____

g Explain why an observer behind the terrapin experiences a lower frequency and longer wavelength when the terrapin is moving compared to when it was stationary. (M)

_____

_____

_____

_____

_____

_____

_____

h Calculate the observed wavelength in front of the terrapin using $\lambda_{in\,front} = v_w T - v_s T$ (check your answer by measurement). (A)

_____

_____

i Calculate the observed frequency in front of the terrapin using $f' = f\dfrac{v_w}{v_w + v_s}$. (A)

_____

_____

j Explain why an observer in front of the terrapin experiences a higher frequency and shorter wavelength when the terrapin is moving compared to when it was stationary. (M)

_____

_____

_____

_____

_____

_____

_____

  ISBN: 9780170368179

Upon seeing some food, the terrapin pushes harder causing it to speed up, but the frequency of its flapping and the speed of the water waves remain constant. The new wave pattern is shown below.

k   By considering the position of the wavefronts and the terrapin after 2.0 s (black), explain why the terrapin must be travelling at 2.0 cm s⁻¹ to produce this pattern. (M)

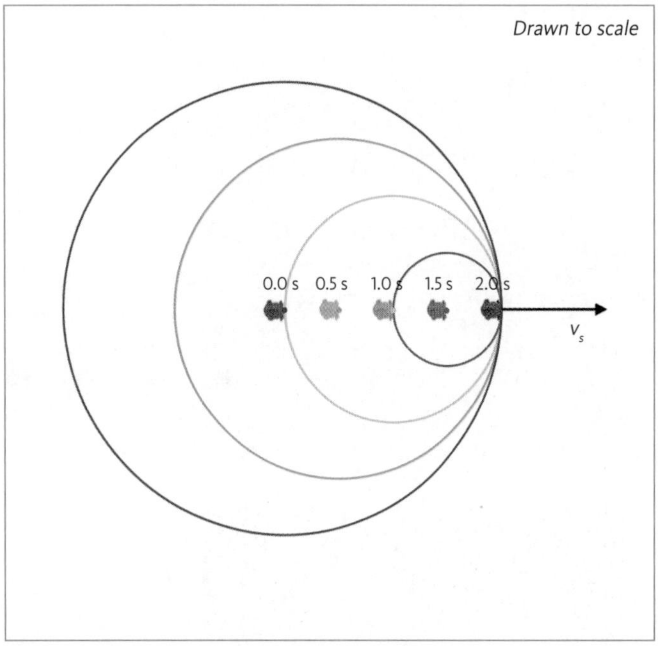

Drawn to scale

0.0 s   0.5 s   1.0 s   1.5 s   2.0 s

v_s

_____

_____

_____

_____

_____

_____

_____

l   Calculate the new observed frequency of the waves behind the terrapin. (A)

_____

_____

m   Explain how increasing the terrapin's swimming speed has affected the observed wavelength and frequency behind the terrapin. (M)

_____

_____

_____

_____

_____

_____

n   Describe the resultant wavefront just in front of the terrapin and explain how this will cause the terrapin problems as it continues to swim forwards. (E)

_____

_____

_____

_____

_____

_____

_____

The terrapin continues to swim at 2.0 m s⁻¹ but enters a shallower region where the water waves traveller slower at only 1.5 m s⁻¹.

o   Complete the diagram below to show the wave pattern surrounding the terrapin after 2.0 seconds.

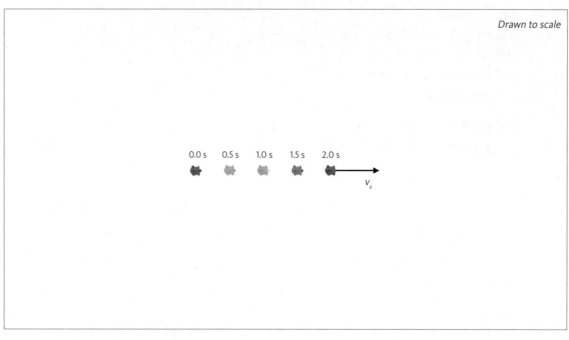

Drawn to scale

2   Wilbur and Orville are brothers and they decide to play with their model airplanes in a local field. When they reach the field, they find their friend Richard has beaten them to it.

Richard's airplane engine **emits a constant sound** of frequency 680 Hz (wavelength 0.500 m) and can fly at a top speed of 23.8 m s⁻¹. As the plane approaches the stationary group, they hear a sound of higher frequency.

Take the speed of sound in air as v_w = 3.40 x 10² m s⁻¹.

a   Show that the observers hear a frequency of 731 Hz as Richard's airplane approaches them. (A)

_____

_____

_____

Richard flies the airplane very low over their heads and the pitch of the note drops suddenly as the plane passes and flies straight away from them.

b   Show that the frequency of the note changes by 95.7 Hz. (M)

_____

_____

_____

   ISBN: 9780170368179

c Draw a labelled graph to show how the frequency changes as the airplane is approaching, passing overhead and then flying away. Use values where possible. (M)

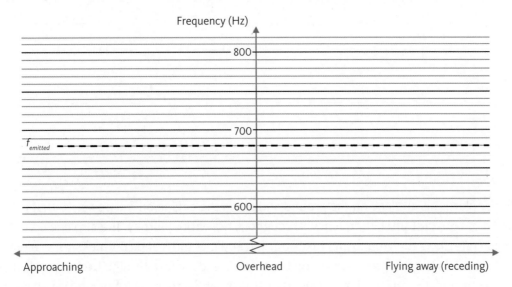

The wavelength of the sound that is heard as the airplane is approaching is 0.465 m, and flying away (receding) is 0.535 m.

d Draw a graph to show how the wavelength changes as the airplane approaches, passes overhead and flies away. Use values where possible. (M)

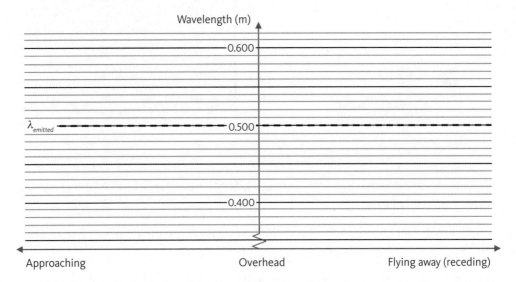

e Compare and contrast the change in frequency and wavelength when approaching and receding and include any calculations. (E)

_____

_____

_____

_____

_____

_____

_____

_____

f Explain why the observers only hear the same pitch as the emitted pitch when the airplane flies directly overhead. (M)

_____

_____

_____

_____

_____

_____

Wilbur places his airplane on the ground and starts the engine on full power so that it emits a sound of **constant pitch** (frequency) throughout the entire journey. The airplane accelerates at a steady rate along the ground and takes off. The airplane turns and continues to accelerate at the same rate in a large 90° arc centred on the three friends before turning back towards them and returning at its top speed. The airplane passes just behind them, as shown in the diagram below.

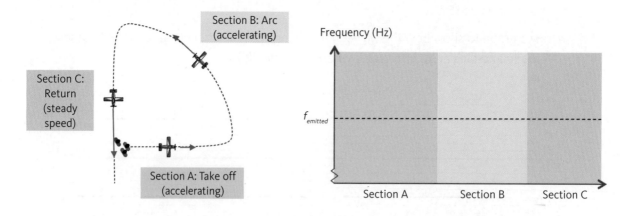

g Sketch a graph to show how the observed pitch (frequency) of the engine changes throughout the journey. (E)

Later in its flight, Wilbur's airplane is travelling away from the observers at a speed of 20.4 m s⁻¹ and they hear it making a sound of frequency 5.00 x 10² Hz.

h Determine the frequency of the sound being emitted by Wilbur's airplane engine. (A)

_____

_____

_____

Orville then launches his airplane and it flies straight towards them. The airplane's engine emits a sound of frequency 6.00×10^2 Hz, but as the airplane is approaching the observers, they hear a frequency of 6.40×10^2 Hz.

i Calculate the speed at which Orville's airplane is travelling towards them. (M)

_____

_____

_____

_____

_____

_____

Wilbur suggest that Orville fits a microphone to his airplane and then measure the Doppler effect while following Richard's aircraft when they are both flying at 20.0 m s⁻¹. Richard predicts that the microphone will record a lower frequency but Wilbur and Orville think that no Doppler shift will be recorded by the microphone.

j Who is right? Explain your reasoning. (E)

_____

_____

_____

_____

_____

_____

_____

_____

_____

3 Burt likes motorcycle racing. He warms up his motorcycle by riding it at a constant speed around a large circular track of radius 47.8 m. A distant observer hears Burt practising and notices that the pitch of the engine rises and falls as Burt approaches and then rides away as shown below. Take the speed of sound as 3.40×10^2 m s⁻¹.

Not to scale

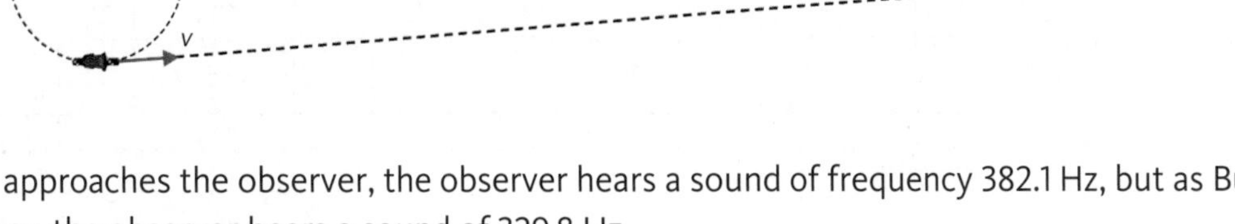

As Burt approaches the observer, the observer hears a sound of frequency 382.1 Hz, but as Burt rides away, the observer hears a sound of frequency 329.8 Hz.

a Show that Burt is riding around the track at a speed of 25.0 m s⁻¹. (E)

_____

_____

_____

_____

_____

_____

_____

_____

b Draw a graph to show how the frequency changes with time as Burt rides around the track at a steady speed. Start the graph when Burt is at the top of the circular track and travelling away from the observer and show at least two complete cycles. You may assume that the observer is sufficiently far away that Burt is at the top and bottom of the circle when the Doppler effect is at a maximum. (E)

_____

_____

_____

_____

_____

_____

_____

_____

_____

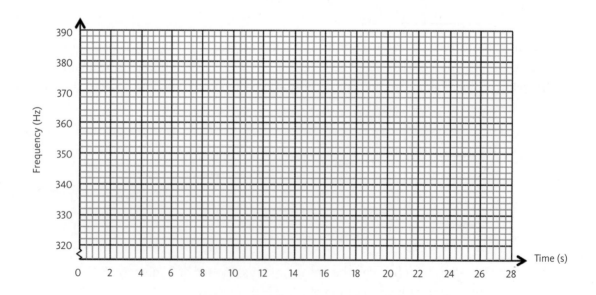

 ISBN: 9780170368179

4   As Arapeta waits for a bus, he notices an airplane passing high overhead. He hears the sound from its engines gradually descend in pitch as it flies overhead at a velocity of 250 m s^{-1} (east). The airplane emits a constant sound of frequency 72.0 Hz and is flying at an altitude of 5.00 km. Take the speed of sound at this altitude as 3.30 x 10^2 m s^{-1}.

   a   Complete the diagrams below to show how the vector components of velocity towards (parallel) and perpendicular to Arapeta change as the airplane flies overhead. Position 1 has been done for you.

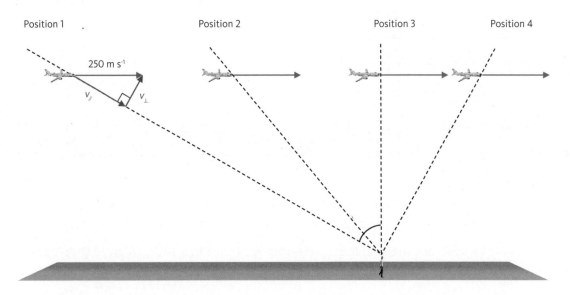

   b   Explain why the perpendicular component of velocity has no effect on the pitch of the sound heard by Arapeta. (M)

   _____

   _____

   _____

   _____

   _____

   c   Explain why the frequency of the sound of the airplane that Arapeta hears gradually decreases as the airplane approaches. (M)

   _____

   _____

   _____

   _____

   _____

   _____

   _____

   _____

   _____

   _____

d With the aid of a vector diagram, show that the wavelength of the sound heard by Arapeta is 1.58 m when the airplane is at Position 1. (M)

_____

_____

_____

_____

_____

_____

_____

_____

Scholarship question

Continuing on from Question 4

The airplane emits a constant sound of frequency 72.0 Hz and continues flying at an altitude of 5.00 km. It is moving at a steady speed of 250 m s⁻¹ as it moves from Position 3 to Position 4.

e Calculate the distance the airplane travels from Position 3 to Position 4 given that Arapeta hears a change in frequency of 19.78 Hz between the two positions.

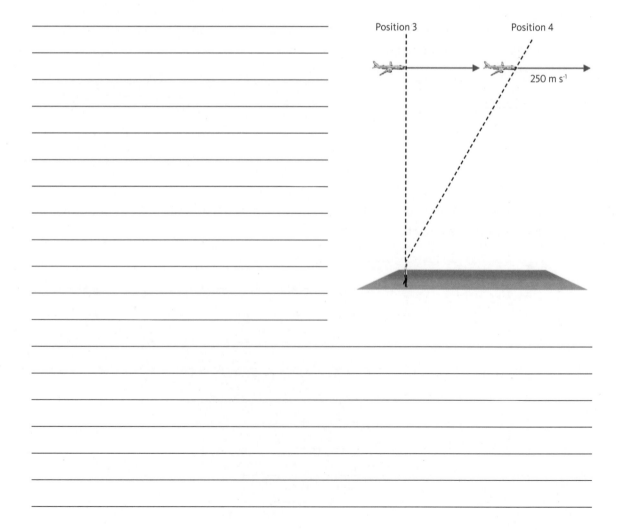

_____

_____

_____

_____

_____

_____

_____

_____

_____

_____

_____

_____

_____

_____

_____

 ISBN: 9780170368179

4

Mechanical systems

Achievement Standard 91524 (P3.4) requires students to demonstrate understanding by connecting concepts or principles that relate to mechanical systems. The standard is worth 6 credits and is assessed externally.

Translational motion

- Centre of mass (one and two dimensions).
- Conservation of momentum and impulse (two dimensions only).

Circular motion and gravity

- Velocity and acceleration of, and resultant force on, objects moving in a circle under the influence of two or more forces.
- Newton's Law of Gravitation.
- Satellite motion.

Rotating systems

- Rotational motion with constant angular acceleration.
- Torque.
- Conservation of angular momentum.
- Rotational inertia.
- Conservation of energy.

Oscillating systems

- The conditions for Simple Harmonic Motion.
- Variation of displacement, velocity and acceleration with time.
- Phasor diagrams.
- Damped and driven systems.
- Conservation of energy.
- Angular frequency.
- Reference circles.
- Resonance.

Relationships

$$d = r\theta \qquad v = r\omega \qquad a = r\alpha \qquad \omega = \frac{\Delta\theta}{\Delta t} \qquad \alpha = \frac{\Delta\omega}{\Delta t} \qquad \omega = 2\pi f \qquad E_{K(rot)} = \frac{1}{2}I\omega^2 \qquad \omega_f = \omega_i + \alpha t$$

$$\theta = \frac{(\omega_f + \omega_i)}{2}t \qquad \omega_f^2 = \omega_i^2 + 2\alpha\theta \qquad \theta = \omega_i t + \frac{1}{2}\alpha t^2 \qquad \tau = I\alpha \qquad L = mvr \qquad L = I\omega \qquad F_g = \frac{GMm}{r^2} \qquad T = 2\pi\sqrt{\frac{l}{g}}$$

$$T = 2\pi\sqrt{\frac{m}{k}} \qquad y = A\sin\omega t \qquad v = A\omega\cos\omega t \qquad a = -A\omega^2\sin\omega t \qquad a = -\omega^2 y \qquad y = A\cos\omega t \qquad v = -A\omega\sin\omega t \qquad a = -A\omega^2\cos\omega t$$

$$x_{CoM} = \frac{m_1 x_1 + m_2 x_2}{m_1 + m_2}$$

4.0 Translational motion

Centre of mass (CoM)

When an external force is applied to an object, it can affect its translational and/or rotational motion. Consider a ruler at rest on a frictionless horizontal desk. Pushing the ruler in the centre will cause the ruler to slide across the desk without any rotation. But if the applied force does not act through the centre of the ruler, then it slides forwards at the same time as spinning about the centre of the ruler.

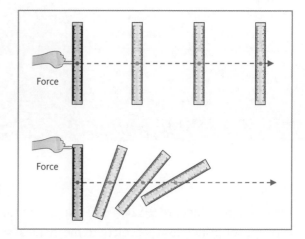

The **centre of mass** of a body is the point through which any applied force **results in only translational motion** and does not cause any rotational motion.

To simplify a lot of complex mechanical problems, this can be expressed as:

The **centre of mass** of a body is the single point at which the entire **mass** of a body can be considered to be **concentrated**.

Centre of gravity

The **centre of gravity** of a body is the single point at which the entire **weight** of a body can be considered to **act**.

In a uniform gravitational field, the centre of mass and centre of gravity coincide.

Stability

The stability of an object depends upon the position of the centre of gravity in relation to the base and the width of the base. Provided the line of action of the weight force acts inside the base, the object will return to a stable position. When the line of action of the weight force acts outside of the base, then the torque causes the object to fall over.

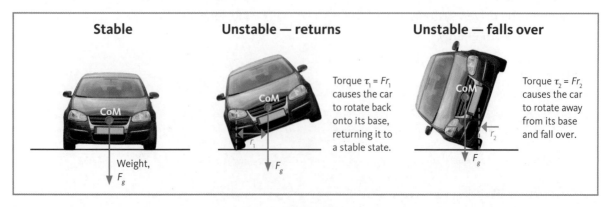

A symmetrical object, such as a rectangular block or sphere, has its centre of mass at the geometrical centre of the object, but for unusual shapes with varying density, the centre of mass can lie towards one end, for example a hammer; or even outside the object, for example a lab stool.

     ISBN: 9780170368179

Centre of mass calculations

The distance of the centre of mass of a system from some chosen point, x_{CoM}, can be found using the formula:

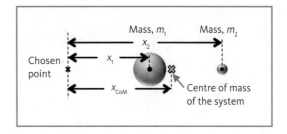

$$x_{CoM} = \frac{m_1 x_1 + m_2 x_2}{m_1 + m_2}$$

For an object made up of more than two masses, this formula can be stated more generally as:

$$x_{CoM} = \frac{\Sigma mx}{\Sigma m}$$

which states:

$$\text{position of the centre of mass (m)} = \frac{\text{sum of the mass (kg) x distance of the separate parts (m)}}{\text{total mass of the system (kg)}}$$

Note: distances are vectors so take displacements to the right as positive.

Problems can be simplified by placing the 'chosen point':

- in the centre of one of the masses (so $x_1 = 0$ or $x_2 = 0$), or
- at the centre of mass of the system (so $x_{CoM} = 0$).

Ratio solutions to centre of mass problems

From the diagram above it can be seen that:

- the centre of mass of a two-body system will always lie on the line joining the centres, and
- the centre of mass of the system is closer to the larger mass (m_1) than the smaller mass (m_2).

This means that centre of mass problems can be quickly solved using **ratios of masses and distances** as measured from the centre of mass of the system.

Ratio solution 1
The ratio of the masses is equal to the inverse ratio of the distances:

$$\frac{m_1}{m_2} = \frac{x_2}{x_1}$$

Ratio solution 2
The ratio of **mass 1** to the **total mass** is equal to the ratio of **distance 2** to the **total separation**:

$$\frac{m_1}{m_1 + m_2} = \frac{x_1}{x_1 + x_2}$$

For example, if m_1 is ⅗ of the total mass, then x_2 will be ⅗ of the total distance between the masses. Consequently, x_1 will be ⅖ of the total distance between the masses.

Worked example: Centre of mass

Joel is riding the log flume at an amusement park. Joel has a mass of 50 kg and the log has a mass of 150 kg. The log is 3.2 m long and Joel sits at the very back of the log with his centre of mass 0.40 m from the end. Determine the position of the centre of mass of the system from the centre of the log.

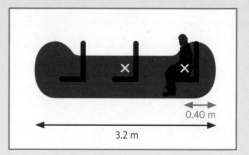

0.40 m

3.2 m

Solution

Given	m_{log} = 150 kg, m_{Joel} = 50 kg, m_{total} = 200 kg
	Distance between the two centres: $d = \dfrac{3.2}{2} - 0.4 = 1.2$ m
	As the distance to the centre of mass is to be measured '**from the centre of the log**', make the 'chosen point' the centre of the log.
Unknown	x_{CoM} = ?
Equations	$x_{CoM} = \dfrac{m_{log}x_{log} + m_{Joel}x_{Joel}}{m_{log} + m_{Joel}}$
Substitute	$x_{CoM} = \dfrac{(150 \times 0) + (50 \times 1.2)}{(150 + 50)}$
Solve	x_{CoM} = 0.30 m (2 sf)

Check using ratios

• The ratio of the masses to the distances gives: $\dfrac{50}{150} = \dfrac{1}{3} = \dfrac{x_{log}}{x_{Joel}}$, so as the CoM of the system is 0.30 m from the CoM of the log, Joel should be 3 x 0.30 = 0.90 m from the CoM of the system. Adding the two distances gives 1.20 m, which is the total distance between Joel and the log.

Or:

• The log is ¾ of the **total mass** of the system, so its centre of mass should be ¼ of the distance between the centre of the log and the centre of Joel, i.e. ¼ × 1.20 = 0.30 m.

Exercise 4A

1   A hammer thrower spins around with the hammer before releasing it. The hammer has a mass of 7.3 kg and its centre of mass is 1.65 m from the centre of mass of the system. The athlete has a mass of 80.3 kg. Determine the distance of the athlete's centre of mass from the centre of mass of the system. (A)

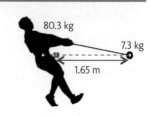

80.3 kg

7.3 kg

1.65 m

_____

_____

_____

_____

2   A desktop toy of a fisherman catching a fish sits precariously on the edge of a desk swinging back and forth. The fisherman has a mass of 48.0 g and the fish has a mass of 72.0 g and they are joined together by a piece of copper wire of negligible mass. The distance between the centre of mass of the fisherman and the fish is 18.0 cm.

   a   Find the distance of the centre of mass of the system from the centre of the mass of the fisherman. (A)

   _____

   _____

   _____

   _____

   _____

   b   Discuss why the fisherman swings but does not fall over when he is tilted to one side and will eventually come to rest in a stable position. (Hint: Consider the torques about the bottom of the pin due to the fisherman and the fish.) (E)

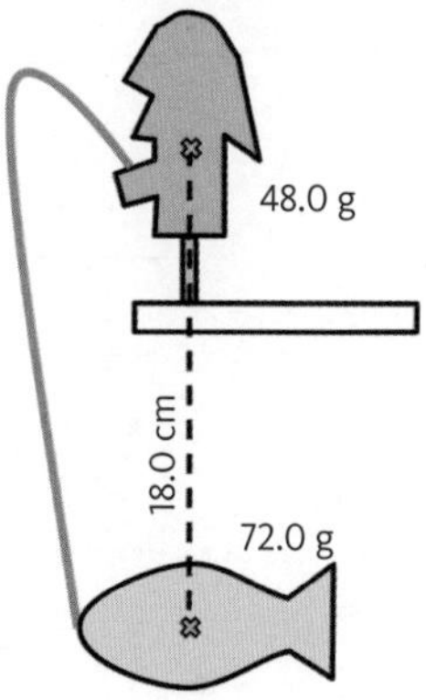

   _____

   _____

   _____

   _____

   _____

   _____

   _____

   _____

3   In July 2015, the New Horizons probe finally reached Pluto. With a mass of 1.305×10^{22} kg, Pluto is so small it was reclassified as a dwarf planet. Its largest moon, Charon, is only slightly smaller than Pluto, so the pair can be considered as a binary planet system that orbits a common centre of mass.

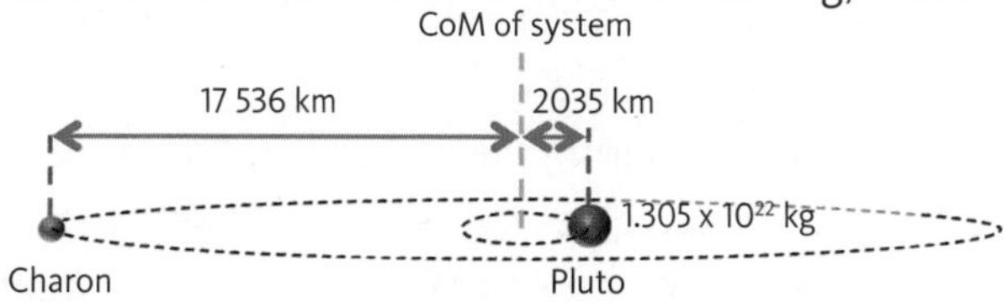

   Determine the mass of Charon given that the orbital radius of Pluto from the common centre of mass is 2035 km, whereas Charon orbits at a distance of 17 536 km. (M)

   _____

   _____

   _____

   _____

   _____

   _____

4 During a motorcycle and sidecar race, the passenger changes position to assist with cornering (see photograph at right). When travelling straight, the passenger crouches down in the sidecar as shown in the diagram below.

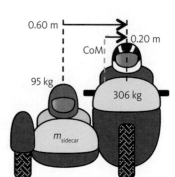

The combined centre of mass of the rider/bike is:
- 0.60 m from the combined centre of mass of the passenger/sidecar, and
- 0.20 m from the centre of mass (CoM) of the whole system.

The rider and bike have a combined mass of M_{RB} = 306 kg and the passenger has a mass of m_p = 95 kg.

a Show that the mass of the sidecar m_{sc} = 58 kg. (M)
(Hint: take the combined CoM of the passenger/sidecar as the chosen point.)

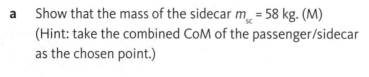

_____

_____

_____

_____

_____

_____

_____

_____

_____

b When cornering, the passenger leans into the bend so his centre of mass is now 50.0 cm from the centre of mass of the sidecar. This causes the centre of mass of the system to move to the left.

By considering the passenger, sidecar, and rider/bike as three parts of a system, determine how far the centre of mass **moves** as a result of the passenger leaning into the bend. (M)

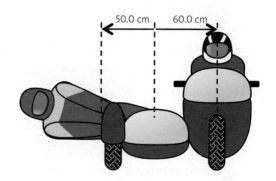

_____

_____

_____

_____

_____

_____

_____

_____

Scholarship questions

5 Following on from question **4** about motorcycle racing, explain why the passenger leans into a bend to improve cornering.

_____

_____

_____

_____

_____

_____

_____

_____

_____

6 During a catamaran race, competitors will stand on the side of the boat to get the maximum speed out of the available wind pushing sideways on the sail. The technique is referred to as 'stacking out' and is shown in the picture and diagram below, where:

m_y = mass of the yachtswomen
M = mass of the catamaran
m_s = mass of the sail and mast
l = length of the mast
CoM = centre of mass

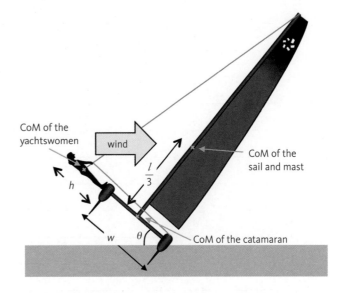

a Show that the horizontal distance of the centre of mass from the centre of mass of the yachtswomen is given by:

$$x_{CoM} = \frac{3(M + m_s)(h + w)\cos\theta + 2lm_s\sin\theta}{6(M + m_s + m_y)}$$

_____

_____

_____

_____

_____

_____

b Using the data given below, determine the horizontal position of the centre of mass when the catamaran is sailing at an angle of 40° to the surface of the water, and hence discuss the conditions necessary for the catamaran to maintain this position during a race. Calculate the average force due to the wind and state any assumptions that you use.

- Combined mass of the yachtswomen $m_y = 140$ kg
- Mass of the catamaran $M = 155$ kg
- Combined mass of the sail and mast $m_s = 14.8$ kg
- Height of the women $h = 1.70$ m
- Height of the mast $l = 9.00$ m
- Width of the catamaran $w = 3.00$ m

_____

_____

_____

_____

_____

_____

_____

_____

_____

_____

_____

_____

_____

_____

_____

_____

_____

_____

_____

_____

_____

_____

Momentum and the centre of mass of a system in one dimension

The **Principle of Conservation of Momentum** states that:

> The total momentum of a system will remain constant throughout an interaction (collision or 'explosion') provided there is no **net** external force acting on the system.

Consider a simple situation of two stationary ice skaters at rest on frictionless ice. If either skater pushes on the other skater, they will drift apart.

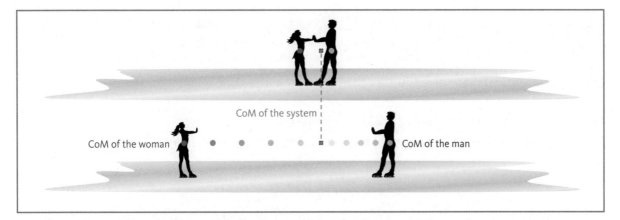

Interactions like this can be explained by thinking about either:

* the motion of the individual parts of the system (i.e. the two skaters), or
* the motion of the centre of mass (CoM) of the system.

Motion of the individual parts of a system

The behaviour of the skaters can be explained using **Newton's Laws of Motion**. When one skater pushes the other skater, both the skaters will experience the same size force according to **Newton's third law**, which states:

> If object A exerts a force on object B, then object B will exert an equal but oppositely directed force on object A.

Hence:

$$F_{\text{man on the woman}} = -F_{\text{woman on the man}}$$

Newton's second law states that:

> The change in momentum, Δp, of an object is directly proportional to the time, t, of an event and the force, F, applied, as $\Delta p = F\Delta t$.

So as the force of the interaction is applied for the **same amount of time** on the man and the woman:

$$\Delta p_{\text{woman}} = -\Delta p_{\text{man}}$$

This states that the change in momentum of the woman will be equal to the change in momentum of the man but in the **opposite direction** (hence the **negative sign**).

As the woman has less mass than the man, the same change in momentum results in a **greater change in her speed**: = ↓ ↑ = ↑ ↓

$$\Delta p_{woman} = m_{woman} \Delta v_{woman} \text{ and } \Delta p_{man} = m_{man} \Delta v_{man}$$

Once they lose contact, they will move at a steady speed away from each other in accordance with **Newton's first law**:

> Every object continues in a state of rest or of uniform (non-accelerated) motion in a straight line, unless acted on by a net external force.

Consequently, the woman travels further than the man in the same amount of time.

Motion of the centre of mass (CoM) of the system

The momentum of the centre of mass of a system is the sum of the momenta of the individual parts:

$$p_{CoM} = \Sigma p_{parts}$$

As momentum is a **vector quantity**, it has **magnitude** and **direction**, so the direction in which the parts move must be considered (by choosing a positive direction, for example + →).

In the example of the skaters above, the initial momentum of the centre of mass was zero, and as the two skaters travel in opposite directions, their total momentum will sum to zero.

Velocity of the centre of mass of a system

As momentum $p = mv$, the equation above can be rewritten as:

$$(m_1 + m_2 + ...)v_{CoM} = m_1 v_1 + m_2 v_2 + ...$$

which can be rearranged to give:

$$v_{CoM} = \frac{m_1 v_1 + m_2 v_2 + ...}{m_1 + m_2 + ...}$$

and expressed as:

$$v_{CoM} = \frac{\Sigma mv}{\Sigma m}$$

which states:

$$\frac{\text{velocity of the centre of mass}}{\text{(m s}^{-1}\text{)}} = \frac{\text{sum of the momenta of the separate parts (kg m s}^{-1}\text{)}}{\text{total mass of the system (kg)}}$$

Worked example: Ariel and Grant skating

Ariel and Grant are skating together on frictionless ice. Grant pushes Ariel ahead of him causing him to slow down to 2.0 m s^{-1} ($\rightarrow$) and Ariel to speed up to 5.0 m s^{-1} ($\rightarrow$) relative to the ice. Ariel has a mass of 56 kg and Grant has a mass of 84 kg.

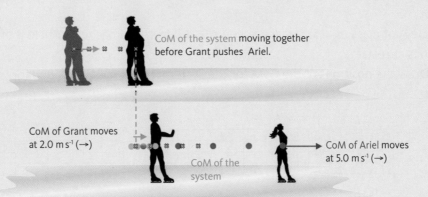

CoM of the system moving together before Grant pushes Ariel.

CoM of Grant moves at 2.0 m s^{-1} ($\rightarrow$)

CoM of the system

CoM of Ariel moves at 5.0 m s^{-1} ($\rightarrow$)

a Calculate the velocity of the centre of mass of the system before the push and explain the physics principles behind your answer.

b Shortly after the push, Grant and Ariel are 7.5 m apart. Determine the distance of Grant from the centre of mass of the system.

Solution

a Grant applies a force onto Ariel causing her momentum to increase, but according to Newton's third law, she applies an equal but opposite force to Grant causing his momentum to decrease. Momentum is conserved as the ice is frictionless, so there is no net external force. The momentum of the centre of mass remains constant throughout the interaction. As the mass of the system remains constant, this means that the velocity of the centre of mass is the same throughout the interaction.

Given	$m_G = 84$ kg, $v_{Gf} = 2.0$ m s^{-1}, $m_A = 56$ kg, $v_{Af} = 5.0$ m s^{-1}
Unknown	$v_{CoM} = ?$
Equations	Conservation of momentum states that: $p_{CoM} = p_{Gf} + p_{Af}$ and $p = mv$
Substitute	$(84 + 56) v_{CoM} = (84 \times 2.0) + (56 \times 5.0)$
	$v_{CoM} = \dfrac{448}{140}$
Solve	$v_{CoM} = 3.2$ m s^{-1} (2 sf) ($\rightarrow$)

b

Given	$m_G = 84$ kg, $m_A = 56$ kg, $d = 7.5$ m
Unknown	$x_{CoM} = ?$ The distance of the centre of mass from Grant.
Equations	$x_{CoM} = \dfrac{m_G x_G + m_A x_A}{m_G + m_A}$ and taking Grant as the chosen point.
Substitute	$x_{CoM} = \dfrac{(84 \times 0) + (56 \times 7.5)}{84 + 56}$
Solve	$x_{CoM} = 3.0$ m (2 sf) ($\rightarrow$) from Grant.

Exercise 4B

In questions **1–3**, two particles are approaching each other. Their positions at different times are shown to scale on a grid, where 1 square = 1 m. Use the information provided to solve for the unknown value.

1 Particle A of mass m_A = 5.0 kg and speed v_A = 4.0 m s^{-1} approaches particle B of mass m_B = 5.0 kg, which is at rest.

    a Draw the positions of the CoM for the system; t = 0 s has been done for you.

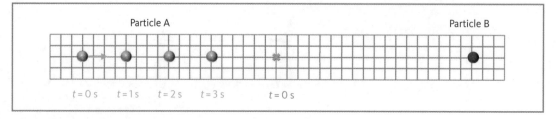

    b Determine the speed of the CoM using the diagram above.

_____

_____

    c What fraction is the mass of particle A to the mass of the whole system?

_____

_____

    d What fraction is the speed of the CoM compared with the speed of particle A?

_____

_____

    e Calculate the speed of the CoM by considering the momentum of the system.

_____

_____

2 Particle E of mass m_E = 5.0 kg and speed v_E = 4.0 m s^{-1} approaches particle F of mass m_F = 15.0 kg, which is at rest. Use a scale of 1 square = 1 m.

    a Draw the positions of the CoM for the system; t = 0 s has been done for you.

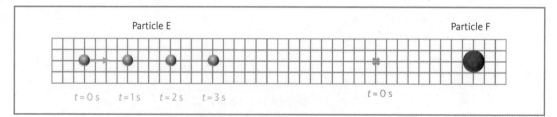

    b Show that the speed of the CoM is 1.0 m s^{-1} using the diagram above.

_____

_____

c Determine how long it will take the centre of particle E to reach the centre of particle F using the formula: $v_E = \dfrac{\Delta d_{E \text{ to } F}}{\Delta t}$

_____

_____

d Determine how long it will take the CoM of the system to reach the centre of particle F using the formula: $v_{CoM} = \dfrac{\Delta d_{CoM \text{ to } F}}{\Delta t}$

_____

_____

3 Particle G of mass $m_G = 5.0$ kg and speed $v_G = 2.0$ m s⁻¹ approaches particle H, which is moving in the opposite direction at speed v_H. The position of the centre of mass has been drawn on the diagram. Use a scale of 1 square = 1 m.

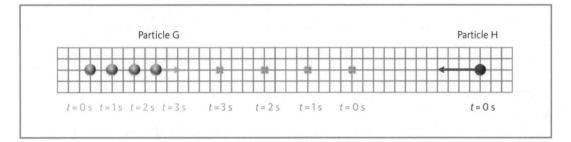

a Using the diagram above, show that the mass of particle H is 10 kg.

_____

_____

_____

b Determine the size and direction of the velocity of particle H.

_____

_____

_____

c Draw the positions of the particle H at $t = 1$, 2 and 3 s; $t = 0$ s has been done for you.

4 Beth and Rachel are performing an ice-dancing routine using a long ribbon. The ice is frictionless and they are moving to the left at a steady speed of 2.0 m s⁻¹ while holding the ribbon tight. Beth is 2.56 m from the centre of mass of the system and has a mass of 56.0 kg. Rachel has a mass of 64.0 kg.

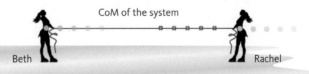

a Calculate the distance between Beth and Rachel.

_____

_____

_____

_____

Rachel pulls the ribbon towards her causing Beth to come to a stop.

Beth   Rachel

b Describe and explain what happens to the motion of both the centre of mass and Rachel.

_____

_____

_____

_____

_____

_____

c Determine the velocity of Rachel after she has applied the force to the ribbon.

_____

_____

_____

5 A 2.00 m long Hamboard (a very long skateboard, which is designed to behave like a surfboard) is rolling along at a steady speed of 2.30 m s^{-1} ($\rightarrow$) when a large cat of mass 3.40 kg drops vertically downwards onto the front of the board and holds on with its claws. This causes the Hamboard to slow down to a speed of 1.84 m s^{-1} ($\rightarrow$).

2.30 m s^{-1} ($\rightarrow$)

a Explain why the velocity of the Hamboard decreases when the cat lands and holds on to the Hamboard where it lands. (E)

_____

_____

_____

_____

_____

_____

_____

b Show that the Hamboard has a mass of 13.6 kg. (E)

_____

_____

_____

_____

_____

_____

_____

The cat sits down with its centre of mass 10.0 cm from the front of the 2.00 m long Hamboard.

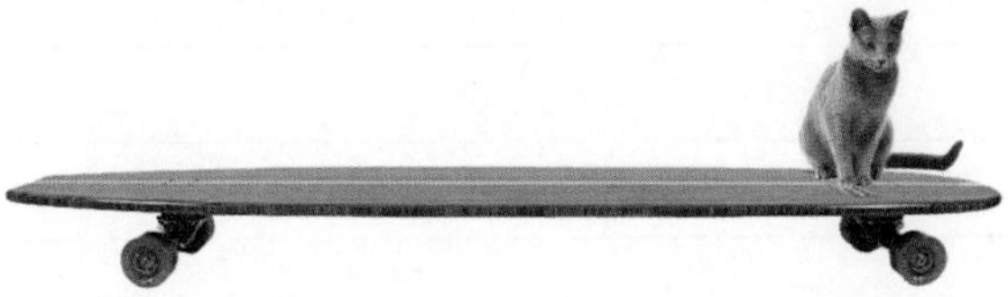

c Determine the distance of the centre of mass of the system from the back of the Hamboard. (A)

_____

_____

_____

_____

_____

While the Hamboard is rolling at a steady speed of 1.84 m s^{-1} ($\rightarrow$) relative to the ground, the cat walks towards the back of the Hamboard at a speed of 1.60 m s^{-1} ($\leftarrow$) relative to the Hamboard and sits down again.

d Show that the velocity of the cat relative to the ground is 0.24 m s^{-1} ($\rightarrow$) as it walks towards the back of the Hamboard. (A)

_____

_____

_____

_____

_____

e   Discuss what will happen to the speed of the centre of mass of the system and the Hamboard relative to the ground when:

i   the cat sets off

ii   walks down the Hamboard, and

iii stops at the other end.

Support your discussion with calculations. (E)

Momentum and the centre of mass of a system in two dimensions

The Principle of Conservation of Momentum applies to interactions in which the separate parts of the system are moving in two and three dimensions. Consider the collision during a game of snooker in which the white cue ball strikes a stationary red ball at an angle.

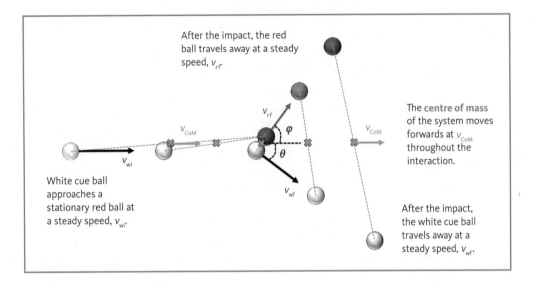

The interaction above can be explained in terms of:

- the **total momentum** of the system, or
- the **impulse** (change in momentum) of part of the system due to the collision.

Momentum of the centre of mass of a system

The momentum of the centre of mass of a system is equal to the sum of the momenta of the separate parts, and the momentum of a system remains the same throughout any interaction provided there is no net external force, so:

$$p_{CoM} = \text{total } p_{initial} = \text{total } p_{final}$$

For the example above:

$$p_{CoM} = p_{white\ initial} + p_{red\ initial} = p_{white\ final} + p_{red\ final}$$

Momentum is a vector quantity, so the total momentum is the **vector sum** of the momenta of the separate parts, and must be solved using a **vector diagram**. For the example above:

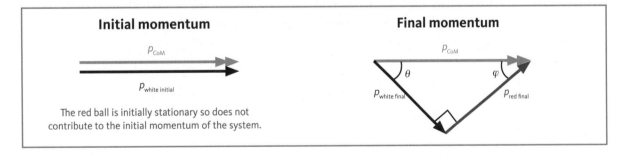

The magnitude and direction of the final vectors can then be found using Pythagoras's theorem and trigonometry for right-angled triangles, or by resolving into components for non-right-angled triangles (*or the sine and cosine rules — however knowledge of these rules is not expected*).

Change in momentum in 2D — impulse

During a collision or 'explosion', internal forces will act between the parts of the system resulting in a change in momentum of the separate parts and can be determined using the impulse formula:

$$\Delta p = F\Delta t$$

which states:

$$\text{impulse (change in momentum) (Ns)} = \text{force (N)} \times \text{time (s)}$$

The change in momentum of each part of the system can also be found as follows:

$$\Delta p = p_{final} - p_{initial}$$

In the example above, the size of the change in momentum of the two balls is given by the formulae:

$$\Delta p_{white} = p_{white\ final} - p_{white\ initial} \quad \text{and} \quad \Delta p_{red} = p_{red\ final} - p_{red\ initial}$$

As the change in momentum involves magnitude and direction, it must be solved using a **vector diagram**. For the example above:

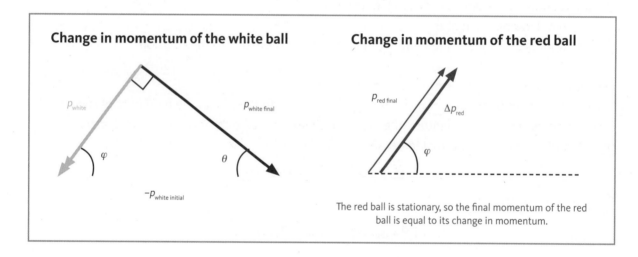

Change in momentum of the white ball

Change in momentum of the red ball

The red ball is stationary, so the final momentum of the red ball is equal to its change in momentum.

The magnitude and direction of the change in momentum can then be found using Pythagoras's theorem and trigonometry or by resolving into components.

As the force of the white ball on the red ball is equal but opposite to the force of the red ball on the white ball (Newton's third law), the change in momentum of the white ball will be equal but opposite to the change in momentum of the red ball, as shown in the diagram above, so:

$$\Delta p_{white} = -\Delta p_{red}$$

Worked example: Air hockey

A single air hockey puck of mass 42.0 g slides at 4.50 m s^{-1} towards a stationary double puck (one puck stacked on top of another). They collide at an angle causing the single puck to travel away at 3.60 m s^{-1} at an angle to its original path. The double puck moves away perpendicular to the direction of the single puck as shown in the diagram. The collision lasts 0.020 s.

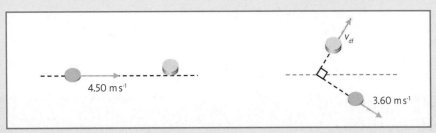

a Calculate the velocity of the centre of mass of the system of puck and double puck before the collision.

b Show that the velocity of the double puck after the collision is 1.35 m s^{-1}.

c Calculate the size of the force exerted by the single puck on the double puck.

Solution

a

Given	$m_s = 0.0420$ kg, $m_d = 0.0840$ kg, $v_{si} = 4.50$ m s^{-1} ($\rightarrow$), $v_{di} = 0.00$ m s^{-1}
Unknown	$v_{CoM} = ?$
Equations	The total momentum of the system is equal to the momentum of the separate parts: $P_{CoM\ initial} = P_{si} + P_{di}$
Substitute	$(0.0420 + 0.0840)v_{CoM} = 0.0420 \times 4.50 + 0.0840 \times 0$
Solve	$v_{CoM} = \dfrac{0.189}{0.126} = 1.50$ m s^{-1} ($\rightarrow$) (3 sf)

b

Given	$m_s = 0.0420$ kg, $m_d = 0.0840$ kg, $v_{si} = 4.50$ m s^{-1} ($\rightarrow$), $v_{sf} = 3.60$ m s^{-1} ($\searrow$)
Unknown	$v_{df} = ?$
Equations	Assuming that there is no net external force, the momentum of the centre of mass of the system remains constant so:

$P_{CoM\ i} = P_{CoM\ f} = P_{sf} + P_{df}$ and $p = mv$.

(As the pucks travel away in different directions, the situation needs to be drawn out to see what is happening then solved using Pythagoras's theorem $P_{CoM\ i}{}^2 = P_{sf}{}^2 + P_{df}{}^2$ and trigonometry.)

Substitute	$P_{CoM\ f} = (0.0420 + 0.0840) \times 1.50$ ($\rightarrow$) $= 0.189$ ($\rightarrow$) N s, and

$P_{sf} = 0.0420 \times 3.60$ ($\searrow$) $= 0.1512$ ($\searrow$) N s

By Pythagoras's theorem, $P_{CoM\ f}{}^2 = P_{sf}{}^2 + P_{df}{}^2$

$0.189^2 = 0.1512^2 + P_{df}{}^2$

$P_{df}{}^2 = 0.01286$

$P_{df} = 0.1134$ N s

Using trigonometry, $\sin\theta = \dfrac{P_{sf}}{P_{CoM\ f}}$

$\theta = \sin^{-1}\left(\dfrac{0.1512}{0.189}\right) = 53.1°$ to the direction of the CoM

Solve $v_{df} = \dfrac{P_{df}}{m_d} = \dfrac{0.1134}{0.0840} = 1.35$ m s^{-1} (3 sf) at 53.1°

c

Given	$p_{di} = 0$ N s, $p_{df} = 0.1134$ N s at $53.1°$, $t = 0.020$ s
Unknown	$F_{s\,on\,d} = ?$
Equations	$\Delta p = F\Delta t$ and $\Delta p = p_f - p_i$
Substitute	$\Delta p = 0.113$ (↗) $- 0$ and so $F = \dfrac{0.1134}{0.020} = 5.67$
Solve	$F_{s\,on\,d} = 5.7$ N (2 sf) at $53.1°$

Exercise 4C

In questions **1** and **2**, two particles form a system. Their positions at different times are shown to scale on a grid where **1 square = 1 m**. Use the information provided to solve for the unknown values. The terms horizontal and vertical have been used to simplify descriptions and both the force due to gravity and friction can be ignored.

1   Particle A of mass $m_A = 3.0$ kg collides with particle B and they stick together as shown in the diagram below. Particle B has a mass of $m_B = 6.0$ kg.

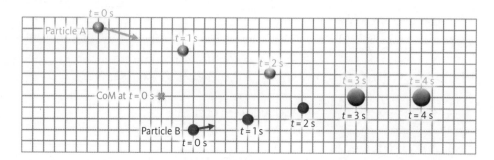

The diagram shows that particle A **moves 2 squares** vertically downwards each second. This means the vertical component of the velocity is $v_{Av} = 2$ m s⁻¹ (↓).

a   Using the grid, show that the vertical component of velocity of particle B is $v_{Bv} = 1$ m s⁻¹ (↑). (A)

_____

_____

b   By considering only the vertical components of **momentum**, calculate the vertical component of the **velocity** of the centre of mass. (A)

_____

_____

_____

c   By considering the horizontal components of momentum of the system, calculate the horizontal component of the velocity of the centre of mass. (A)

_____

_____

_____

   ISBN: 9780170368179

d Mark the positions of the CoM for the system on the diagram above when $t = 1, 2, 3$ and 4 s ($t = 0$ s has been done for you). Label each CoM position with a time. Connect the centres of the masses and the CoM together with a single line for $t = 0, 1$ and 2 s. (A)

2 A system made up of two particles C and D joined together has a mass of $m = 8.0$ kg. The centre of mass of each particle lies at the centre of the system. It is travelling in a straight line at 7 m s^{-1} when it explodes pushing particle C upwards and particle D downwards as shown in the diagram below. The position of the centre of mass of the system is shown throughout the interaction.

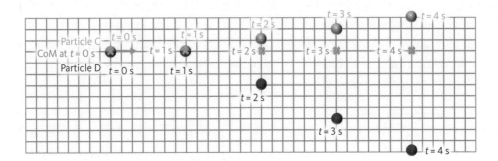

a Using the diagram, determine the vertical components of velocity of particles C and D. (A)

_____

_____

_____

b Explain why the momentum of the centre of mass remains constant despite the explosive force acting on the system. (A)

_____

_____

_____

c By considering the vertical components of the momentum of the two particles and the system, show that $m_C = 6.0$ kg and $m_D = 2.0$ kg. (A)

_____

_____

_____

_____

Particles C and D are both still travelling horizontally at 7.0 m s^{-1}.

d By combining the velocity components for particle C after the explosion, show that the momentum of particle C is $p_C = 42$ kg m s^{-1} at an angle of 8.1°. (M)

_____

_____

_____

_____

e By combining the velocity components for particle D after the explosion, show that the momentum of particle D is $p_D = 15$ kg m s^{-1} at an angle of 23°. (M)

_____

_____

_____

_____

Before the explosion, the size of the total momentum of the system was $p_{CoM} = 56$ kg m s^{-1}.

f By considering the momentum of the system before the explosion and the momentum of particles C and D after the explosion (given above in parts **d** and **e**), discuss whether momentum has been conserved during the explosion. Draw a momentum vector diagram to support your discussion. (E)

_____

_____

_____

_____

_____

_____

_____

_____

_____

_____

_____

_____

_____

_____

3 A red air-hockey mallet of mass 50.0 g slides towards a blue air-hockey mallet of mass 60.0 g, which is at rest.

They collide at an angle causing the red mallet to travel away perpendicular to the direction of the blue mallet as shown in the diagram.

The centre of mass of the system of mallets has a momentum of 0.075 kg m s^{-1}.

a Show that the red mallet is travelling at 1.5 m s^{-1} before the collision. (M)

_____

_____

_____

After the collision, the blue mallet moves away at 1.20 m s^{-1}. The red mallet moves perpendicular to the direction of the blue mallet.

b Determine the momentum of the centre of mass of th system after the collision. (A)

_____

_____

_____

c Show that the momentum of the blue mallet after the collision is 0.072 kg m s^{-1}. (A)

_____

d Draw a vector diagram to show the relationship between the momentum of the centre of mass of the system and the momentum of the two mallets after the collision.

e Calculate the direction of the blue mallet relative to the direction of the centre of mass. (A)

_____

_____

_____

f Calculate the speed and direction of the red mallet after the collision. (M)

_____

_____

_____

_____

_____

The collision lasts 0.030 s resulting in a change in momentum of both mallets.

g   Discuss the relationship between the impulse of the red mallet and the impulse of the blue mallet. Support your answer with calculations and a vector diagram. (E)

_____

_____

_____

_____

_____

_____

_____

_____

h   Calculate the size of the force exerted by the red mallet on the blue mallet. (A)

_____

_____

4   An astronaut is wearing a Manned Manoeuvring Unit (MMU) to help her move around in space. She is travelling at a steady speed when she activates the MMU to make a course change so that she moves towards the entry hatch of a space station at a slower speed. The astronaut, MMU and propulsion gases have a total mass of 204 kg. When she activates the MMU it releases a 4.00 kg jet of nitrogen gas at high speed causing her to change course by 36.87° and move with a momentum of 60.0 kg m s^{-1}. The gas jet travels away with a momentum of 111 kg m s^{-1}.

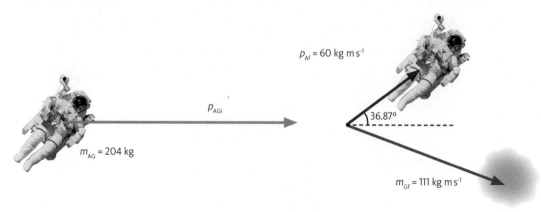

$p_{Af} = 60$ kg m s^{-1}

p_{AGi}

36.87°

$m_{AG} = 204$ kg

$m_{Gf} = 111$ kg m s^{-1}

a Show that the astronaut's vertical component of momentum is 36 kg m s^{-1} as a result of activating the MMU. (A)

_____

_____

b Explain why the vertical component of momentum of the propulsion gases is the same size as the astronaut's. (M)

_____

_____

_____

_____

_____

c Calculate the angle at which the propulsion gases are travelling compared to the original direction of the astronaut. (A)

_____

_____

d Draw a labelled vector diagram to show the relationship between the initial momentum of the system and the final momentum of the astronaut and the propulsion gases. (M)

e Show that the initial speed of the system before the astronaut activates the MMU is 0.75 m s^{-1}. _Hint: This is not a right-angled triangle._ (E)

_____

_____

_____

_____

_____

f Calculate the initial momentum of the astronaut. (A)

_____

_____

g Draw a labelled vector diagram to show the **change in momentum** of the astronaut. (M)

To make the course change, the MMU provided an average force of 49.2 N on the astronaut.

h Determine the time for which the MMU was activated. (M)

Scholarship question

5 The main rotor blades of a helicopter sweep out an area of radius 4.0 m and push air downwards with a speed of 15 m s^{-1} away from the blades as shown in the diagram.

By tilting forwards 20°, the helicopter stops rising and starts to accelerate horizontally at 3.57 m s^{-2}. The density of air is ρ_{air} = 1.3 kg m^{-3}.

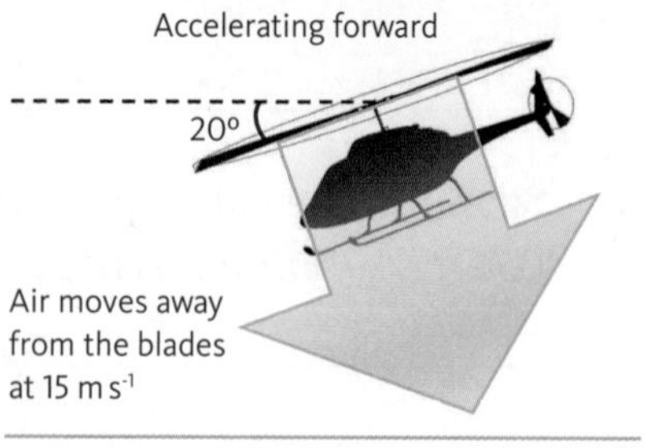

Determine the mass of the helicopter.

Accelerating forward

20°

Air moves away from the blades at 15 m s^{-1}

_____

_____

_____

_____

_____

_____

_____

_____

_____

_____

_____

_____

_____

_____

_____

_____

_____

4.1 Circular motion

Circular motion fundamentals

Centripetal acceleration

An object that is travelling at a steady speed along a circular path must be continually accelerating towards the centre of the circle as its direction and hence velocity is changing. This is known as **centripetal acceleration**, a_c, and can be calculated using:

$$a_c = \frac{v^2}{r}$$ which states: $$\text{centripetal acceleration (ms}^{-2}) = \frac{\text{speed}^2 \text{ (ms}^{-1})^2}{\text{radius (m)}}$$

Centripetal force

According to Newton's second law, a **net force**, F_{net}, is required to cause an object of **mass m** to **accelerate** with acceleration a in the direction of the force, $F_{net} = ma$, hence:

$$F_c = \frac{mv^2}{r}$$ which states: $$\text{centripetal force (N)} = \frac{\text{mass (kg) x speed}^2 \text{ (m s}^{-2})}{\text{radius (m)}}$$

The centripetal force is often the **result of several forces** acting on an object and acts at right angles to the direction of motion, so it has no effect on the speed of the object, but will cause it to change direction.

Horizontal circular motion

On Earth, objects moving along a horizontal circular path will experience the force due to gravity acting perpendicular to the centripetal force. The centripetal force must therefore be provided by another force such as friction, the reaction force from a surface, lift or tension.

Friction on a flat bend E.g. car driving around a bend. This idea is considered in L2 physics.	**Banked corners** E.g. bends of roads, race tracks, velodromes or when aircraft make a banked turn.	**Conical pendulum** E.g. swingball games, amusement park swings, steam engine regulators.

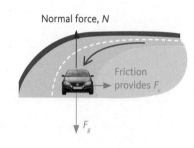

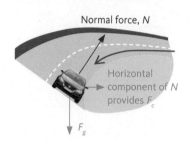

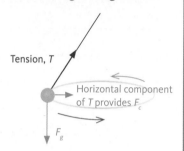

Banked corners

A centripetal force is required to make a vehicle travel around a bend along a semi-circular path.

If the road is flat, then only friction can provide the necessary horizontal force to cause the change in direction.

Wet conditions, poor-quality tyres or oil on the road can all result in the friction force decreasing and affect the ability of vehicles to complete a bend safely.

But if the road is banked at an angle, (θ) the horizontal for a bend of radius r, then the normal force N from the road will provide:

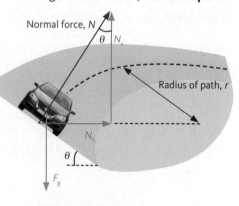

- a vertical support component, N_v,
- a horizontal centripetal component, N_h.

At this speed the vehicle is not reliant on friction at all to complete the bend.

When travelling around a bend at the **ideal speed**, a vehicle travels along a '**horizontal**', '**circular**' path.

- **'Horizontal'** — so vertically the weight force must be balanced by the vertical component of the normal force, i.e.

$$N_v = F_g$$

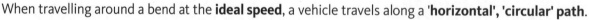

- **'Circular'** — the horizontal component of the normal force provides a centripetal force, i.e.

$$N_h = F_c$$

and as $\tan \theta = \dfrac{N_h}{N_v}$

$$F_c = F_g \tan \theta$$

- Substituting in $F_c = \dfrac{mv^2}{r}$ and $F_g = mg$ and rearranging gives: $\boxed{v_{ideal} = \sqrt{gr \tan \theta}}$

Aircraft banking

When an airplane makes a turn it must roll (tilt) to a banked position so that the lift force, F_L, from the wings can provide a centripetal force as well as maintain the airplane in level flight.

By balancing the vertical forces and equating the components of the lift force it shows that $v = \sqrt{gr \tan \theta}$ as for banked corners. Rearranging this gives the banking angle for an airplane flying with an airspeed v:

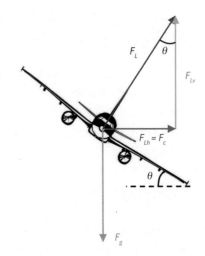

$$\tan \theta = \frac{v^2}{gr}$$

Conical pendulum

A conical pendulum is made up of a bob of mass m suspended from a fixed point by a light string of length l. The bob is pulled to one side through an angle θ and then pushed tangentially so that it moves in a circle of radius r.

In the absence of friction the bob moves in a **'horizontal' 'circle'**.

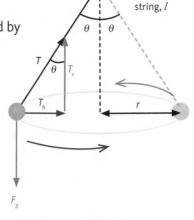

- **'Horizontal'** — so vertically the weight force must be balanced by the vertical component of the tension force in the string, i.e.

$$T_v = F_g$$

- **'Circle'** — the horizontal component of the tension force provides a centripetal force, i.e.

$$T_h = F_c$$

and as $\tan \theta = \dfrac{T_h}{T_v}$

$$F_c = F_g \tan \theta$$

- Substituting in $F_c = \dfrac{mv^2}{r}$ and $F_g = mg$ and rearranging gives: $\boxed{v_{ideal} = \sqrt{gr \tan \theta}}$

This shows that the speed of the conical pendulum is dependent on the radius of the circular path it takes and the angle of the string but is **independent of the mass** of the bob.

Time period

The conical pendulum has been used in clocks because of its regular periodic motion. The time period, T, of a conical pendulum depends upon the circumference of its path, $C = 2\pi r$, and the speed of rotation, $v = \sqrt{gr \tan \theta}$, as $v = \dfrac{C}{T}$, so

$$T = \frac{2\pi r}{\sqrt{gr \tan \theta}}$$

This can be combined with the length of the string, l, to give: $\boxed{T = 2\pi \sqrt{\dfrac{l \cos \theta}{g}}}$

which shows that the time period of the conical pendulum is dependent upon the length of the string and the angle at which it rotates.

(Note: Neither of these formulae is provided and so they must be derived to solve conical pendulum problems.)

Worked example: Going round the bend

A car drives around a long bend on SH 27 near Matamata at the recommended speed of 85 km h⁻¹ (23.61 m s⁻¹). The bend has a radius of 300 m and is banked towards the centre of the curve. Acceleration due to gravity is 9.81 m s⁻².

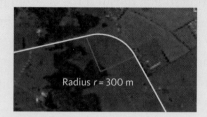

Radius $r = 300$ m

The engine force and friction forces are balanced so the car travels around the bend at a constant speed.

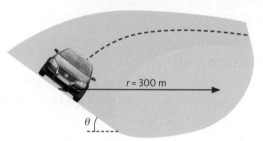

$r = 300$ m

θ

a Draw a labelled free body force diagram to show the other forces acting on the car while driving around the bend.

b Calculate the angle of the bank required to allow the driver to drive around the bend without relying upon friction to make the turn.

c Explain what would happen to the recommended speed of the bend if the radius of the turn were increased to 450 m and drivers are still not relying on friction.

Solution

a The only other forces acting on the car are weight and the normal force from the road.

Note 1: The centripetal force is a component of the normal force so is not shown as a separate force.

Note 2: The normal force cannot be called a support force as it is also causing the car to accelerate towards the centre of the curve.

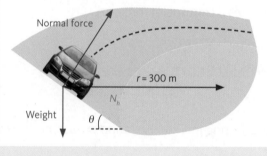

Normal force

Weight

N_h

$r = 300$ m

θ

b

Given	$v = 23.61$ m s⁻¹, $r = 300$ m, $g = 9.81$ m s⁻²
Unknown	Angle of the bank, θ
Equations	$F_c = \dfrac{mv^2}{r}$ and $F_g = mg$ and from the components of the normal force $F_c = F_g \tan \theta$
Substitute	$\tan \theta = \dfrac{23.61^2}{9.81 \times 300} = 0.189$
Solve	$\theta = \tan^{-1} 0.189 = 10.73° = 11°$ (2 sf)

 ISBN: 9780170368179

c

Given	$r = \uparrow$ to 450 m, which is 1.5x the original radius, θ remains constant
Unknown	Speed of the bend, $v = ?$
Equations	From the above proof $v = \sqrt{gr\tan\theta}$ and $F_c = \dfrac{mv^2}{r}$ $\uparrow\sqrt{1.5}^2$
	$\uparrow \quad = \uparrow \quad = \quad = \quad \uparrow 1.5$
Substitute	For the same angle of banking, increasing the radius to 1.5x its original size will increase the maximum speed by $\sqrt{1.5}$.
Solve	As a result of increasing the radius by 1.5x, the ideal speed will increase by $\sqrt{1.5}$, however the centripetal force will remain the same as, as $F_c = \dfrac{m(\sqrt{1.5}\times v)^2}{1.5\times r}$. This means that vehicles will be able to drive at speeds up to: $v = 85 \times$ 1.5 = 104 km h⁻¹ without relying on friction to make the turn.

Extension concepts for scholarship — Friction

The **first law of friction** states that:

> **When two surfaces are in contact, the frictional force between them opposes the motion of one of the surfaces relative to the other.**

Even the flattest, smoothest surfaces are found to be covered in bumps and ridges when viewed at the atomic level. When two surfaces are placed in contact, these bumps and ridges become cold-welded together due to the extremely high pressure applied at these tiny points. These cold welds have to be broken before one surface can move over the other.

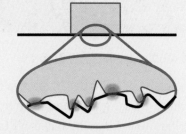

Micro welds between surfaces oppose the motion of surfaces relative to each other.

The **second law of friction** states that:

> **Frictional forces are independent of the area of contact between the surfaces.**

The bumps and ridges on the surfaces mean that the actual area of two surfaces that are touching is actually a lot less (~1/10 000th) than the area of the surfaces that are apparently 'in contact'.

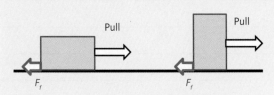

F_f Pull   Pull F_f

Surface area has no effect on the size of the friction force.

Decreasing the area for the same weight reduces the number of points in contact but increases the pressure on each of those points. This flattens the points and ridges and increases the size of the welds; as a result the friction force remains the same regardless of the apparent contact area.

ISBN: 9780170368179

The **third law of friction** states that:

frictional force (N) = coefficient of friction x normal contact force (N)

which can be expressed as:

$$F = \mu N$$

The coefficient of friction, μ, is a constant for a particular pair of surfaces and has a range of values typically between 0 (frictionless) and about 1.

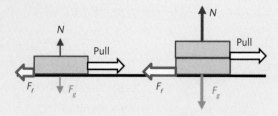

Increasing the normal force increases the friction force.

Increasing the force with which two surfaces are pressed together flattens the bumps and ridges and increases the size of the welds and so increases the frictional force.

Static and kinetic coefficients of friction

For the same two surfaces, the coefficient of static friction (no relative motion) is usually greater than the coefficient of kinetic friction (relative motion between the surfaces). This is because the welds have time to form strong bonds between static surfaces. Note: A rolling object, for example a wheel, experiences static friction because there is no lateral motion unless the object is slipping!

Exercise 4D

Where required, take $g = 9.81 \text{ m s}^{-2}$.

1 The game of swingball involves a tennis ball with a mass of 58.0 g attached to an upright pole by a string 1.10 m long. The ball is hit tangentially so that it moves in an anticlockwise horizontal circle at constant speed. The string makes an angle of 30.0° with the horizontal as shown in the diagram.

    a Draw a labelled free body force diagram on the picture to show the forces acting on the tennis ball and explain how the tennis ball is able to travel in a horizontal circle. (A)

_____

_____

_____

_____

    b Calculate the tension force in the string. (A)

_____

    c Show that the centripetal force acting on the ball is 1.00 N. (A)

_____

d Show that the radius of the path taken by the ball is 0.95 m. (A)

_____

e Show that the speed of the ball is 4.05 m s⁻¹. (A)

_____

_____

f Calculate the time period of the ball. (A)

_____

The string on the swingball game snaps and is replaced with a longer string. When the ball is hit, it travels around the upright pole at the same angle of 30.0°.

g Discuss what effect changing the length will have on the radius, speed, centripetal force and time period of the ball. (E)

_____

_____

_____

_____

_____

_____

_____

2 A swing in a playground is made out of a large tyre of mass 52 kg attached by a chain of negligible mass to a fixed pivot. By pulling the tyre out to one side and pushing it tangentially, the swing moves in a circular path of radius 0.80 m. The swing takes 15 s to complete five revolutions.

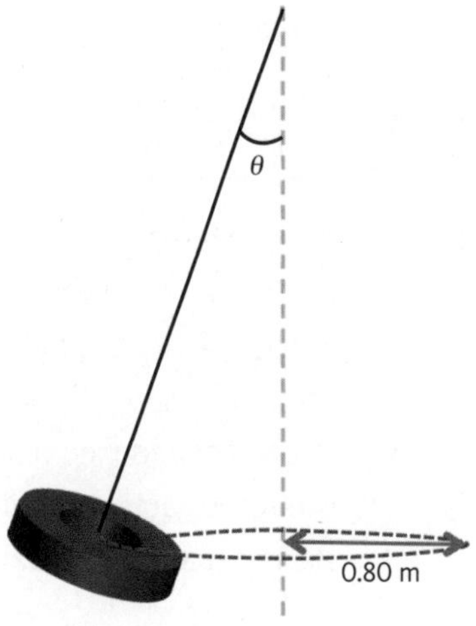

a Show that the time period of the swing is 3.0 s. (A)

_____

_____

b Calculate the centripetal force acting on the tyre. (M)

_____

_____

_____

θ

0.80 m

c Draw a labelled free body force diagram opposite to show:

   i the tension force, T

   ii the weight force, F_g

   iii the net force, F_{net}

  on the tyre.

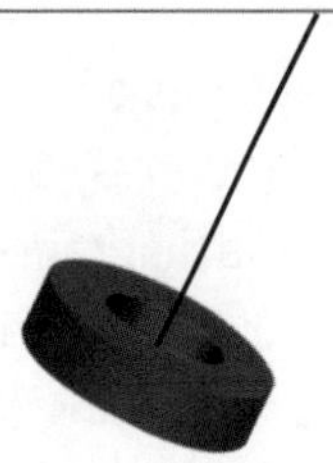

d Calculate the angle about which the tyre is swinging and hence determine the distance from the pivot to the tyre. (M)

_____

_____

_____

_____

_____

A child of mass 26 kg sits in the centre of the tyre and is then pushed so that it travels around the same circular path as before.

e Discuss what effect the child will have on the centripetal force, speed, and time period of the tyre swing. (E)

_____

_____

_____

_____

_____

_____

_____

_____

_____

_____

_____

3 The Bede BD-5 Microjet is one of the smallest jet aircraft in the world with a mass of 530 kg when fully loaded. The jet is flying at a steady speed in a straight line.

a Draw labelled free body force vectors to show all the forces acting when the Microjet is flying at a steady speed in a straight line.

The jet executes a horizontal turn while flying at 216 km h⁻¹ (60 m s⁻¹) by banking at an angle of 40° as shown in the diagram.

b On the diagram, draw labelled free body force vectors showing the lift and weight forces acting on the jet when executing the turn. (A)

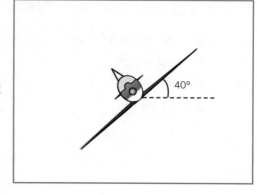

40°

 ISBN: 9780170368179

c Compare and contrast the weight and lift force when flying straight and horizontal to when making a horizontal turn. Support your argument with calculations and state any assumptions you have made. (E)

_____

_____

_____

_____

_____

_____

_____

_____

_____

d Calculate the radius of the horizontal turn. (M)

_____

_____

_____

e Calculate the net acceleration on the jet and hence discuss how the pilot will feel while making the turn. (E)

_____

_____

_____

_____

_____

_____

_____

_____

_____

Scholarship question

4   The ideal speed, v_{ideal}, is the speed at which a car can travel around a bend without relying on friction. However, when a racing car of mass m drives at the maximum speed possible, v_{max}, around a horizontal semi-circular curve of radius r which is banked at an angle θ, the driver relies on the friction between the tyres and the road to maintain this circular path without the car skidding. The size of the friction force, F_f, on the car is:

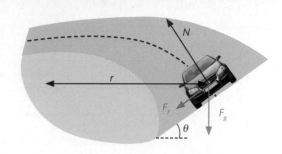

$$F_f = \mu N$$

where N is the normal force on the car from the road surface and μ is the coefficient of static friction of the surface and is a constant for the tarmac and tyres unless they start to slide.

a   Prove that the maximum speed at which the car can complete the bend without sliding is:

$$v_{max} = \sqrt{\frac{gr(\sin \theta + \mu \cos \theta)}{\cos \theta - \mu \sin \theta}}$$

_____

_____

_____

_____

_____

_____

_____

_____

_____

_____

_____

_____

The road is banked at an angle of 30° and the coefficient of static friction $\mu = 0.9$.

b   Show that the maximum speed is 2.3 times greater than the ideal speed at which the car can take the bend without relying on friction and discuss if the additional risk will have a significant effect on the outcome of the race.

_____

_____

_____

_____

_____

_____

   ISBN: 9780170368179

_____

_____

_____

_____

_____

_____

_____

_____

_____

_____

An alternative way of increasing the speed around a bend is for the driver to take a racing line around the bend rather than follow the white central path as shown in the diagram.

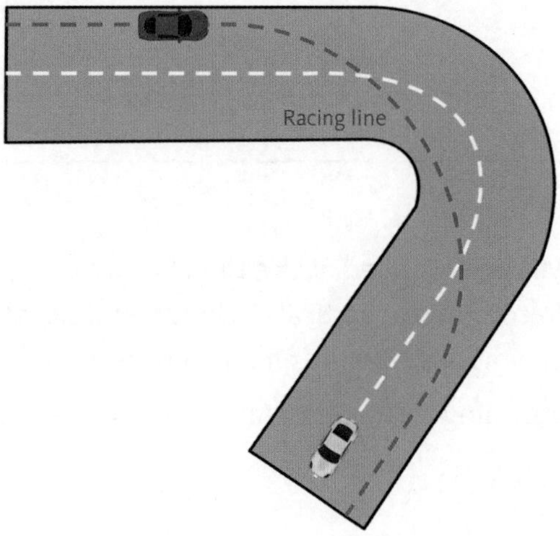

c   Explain why the racing line increases the speed at which the car can safely drive around the bend without relying on friction. Determine the radius of the racing line compared with the central path, that would allow the car to safely complete the bend at maximum speed.

_____

_____

_____

_____

_____

_____

_____

Vertical circular motion

Vertical circular motion can be explained by considering the **forces acting** on the system and the **energy changes** taking place. Objects moving along a vertical circular path will experience a **component** of the force due to gravity acting towards the centre of the circle, which contributes to the centripetal force (see the diagrams below).

Consider a bucket swinging around a vertical circle on the end of a rope. The centripetal force is the net force due to the tension and the component of weight, so as the bucket moves around the circle the tension force changes in size as the component of weight changes.

At the bottom

Weight acts away from the centre so the tension force is at a maximum as it is:

- supporting the weight, and
- providing the centripetal force.

Approaching the top

A component of the weight acts towards the centre and provides some of the centripetal force so the tension force decreases. The tangential component F_{gT} of weight acts to slow the bucket down.

At the top

The weight acts towards the centre and provides most of the centripetal force so the tension force is at a minimum.

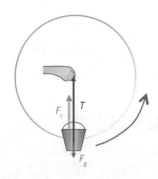

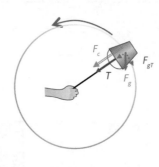

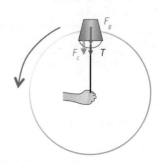

Minimum speed at the top of a circle

To complete a vertical circle successfully an object must be moving at, or greater than, a critical minimum speed. This minimum speed occurs when the weight of the object is the only force providing the centripetal force, $F_c = F_g$ and so $\dfrac{mv_{min}^2}{r} = mg$ hence:

$$v_{min} = \sqrt{gr}$$

At slower speeds the object will no longer follow a circular path but will instead start to fall like a projectile. Consider the bucket on the rope at the top of the circle.

Faster than the minimum speed

The centripetal force is provided by the weight and the tension.

$V > V_{min}$

Circular motion

Minimum speed

The centripetal force is provided by the weight ONLY.

$V > V_{min}$

Circular motion

Rope is loose as there is no tension.

Slower than the minimum speed

The weight force is greater than the centripetal force.

$V > V_{min}$

Parabolic motion

   ISBN: 9780170368179

Vertical circular motion at a constant speed — the loop-the-loop

Airplanes can perform loop-the-loop manoeuvres but at a constant speed as they have the ability to change the engine force and lift from the wings as they travel around the loop. If the airplane is moving at a constant speed, then the centripetal force must be constant, as $F_c = \dfrac{mv^2}{r}$.

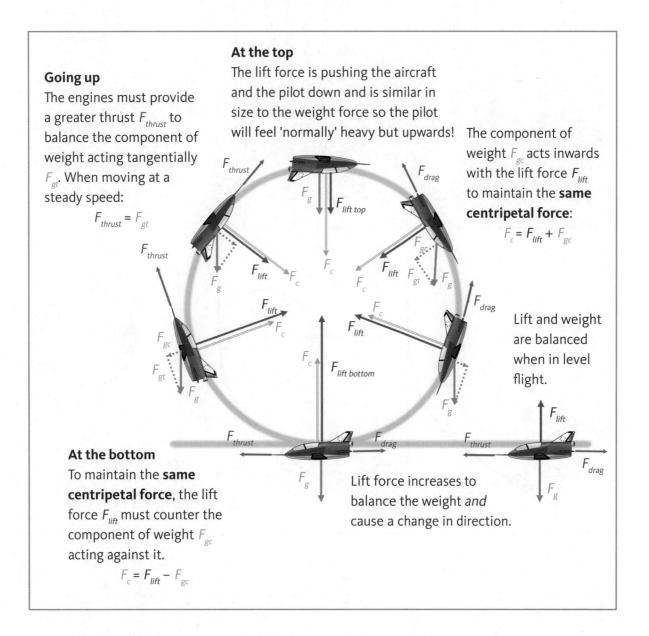

Going up

The engines must provide a greater thrust F_{thrust} to balance the component of weight acting tangentially F_{gt}. When moving at a steady speed:

$$F_{thrust} = F_{gt}$$

At the top

The lift force is pushing the aircraft and the pilot down and is similar in size to the weight force so the pilot will feel 'normally' heavy but upwards!

The component of weight F_{gc} acts inwards with the lift force F_{lift} to maintain the **same centripetal force**:

$$F_c = F_{lift} + F_{gc}$$

Lift and weight are balanced when in level flight.

At the bottom

To maintain the **same centripetal force**, the lift force F_{lift} must counter the component of weight F_{gc} acting against it.

$$F_c = F_{lift} - F_{gc}$$

Lift force increases to balance the weight *and* cause a change in direction.

The lift force from the wings pushing the airplane and pilot up is about three times the size of the weight force, so the pilot will feel very heavy at the bottom.

Forces and vertical circles

Top of the slope

The weight and track force produce a net force down the slope, causing the cart to accelerate.

Bottom of the loop

The track has to supply a force to support the weight of the cart *and* cause a change in direction, so the net (centripetal) force is:

$$F_c = F_{tb} - F_g$$

Hence the force from the track at the bottom is:

$$\mathbf{F_{tb} = F_c + mg}$$

The track force provides passengers with the feeling of heaviness. This is due to the additional centripetal component, making passengers feel heavier at the bottom.

Top of the loop

The weight acts towards the centre so provides some or all of the net (centripetal) force:

$$F_c = F_{tt} + F_g$$

Hence the force from the track at the top is:

$$\mathbf{F_{tt} = F_c - mg}$$

As $F_c = \dfrac{mv^2}{r}$, the feelings of the passengers depend upon the speed of the rollercoaster.

- Fast: If $F_c = 2mg$, normal 'weight' but upwards.
- Minimum speed: $F_c = mg$ so $v_{min} = \sqrt{gr}$ and passengers feel **'weightless'** as $F_{tt} = 0$ N.
- Too slow: $F_c = mg$, cart **falls off** the track.

Vertical circular motion and changing speed

Top of the slope

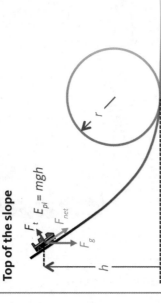

Bottom of the loop

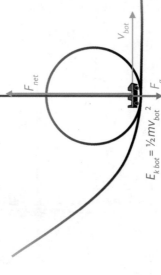

Top of the loop

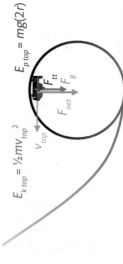

Energy and vertical circles

Top of the slope

The cart is initially stationary so has only gravitational potential energy:

$$E_{pi} = mgh$$

Bottom of the loop

As the cart moves down the track it transforms E_p into E_k, until it reaches the bottom when:

$$E_{pi} = E_{k\,bot}$$

So the cart is moving at:

$$\mathbf{v_{bot} = \sqrt{2gh}}$$

Top of the loop

As the cart rises up the slope, E_k is transferred back into E_p. As the cart must continue moving over the top of the loop at a **minimum speed**, the loop must be lower than the height of the slope.

By the Law of Conservation of Energy, the total energy must remain the same so the total energy at the top of the loop must equal the total energy at the start.

$$E_{total} = E_{pi} = E_{k\,top} + E_{p\,top}$$

Hence:

$$mgh = \tfrac{1}{2}mv_{min}^2 + mg(2r)$$

which combines with the forces ($\mathbf{v_{min} = \sqrt{gr}}$) explanation to give a minimum speed at the top of:

$$\mathbf{v_{top} = \sqrt{\tfrac{2}{5}gh}}$$

ISBN: 9780170368179

Worked example: Cyclist loop-the-loop

In November 2014, Danny MacAskill, world-renowned trials cyclist, successfully cycled around a loop-the-loop. Danny started from rest at the top of a 7.0 m high ramp and freewheeled down to the loop. For simplicity you may assume that the mass of the system is concentrated in the centre of Danny, which moves around a loop of radius 2.00 m.

Take the mass of Danny as 78 kg, his bike as 14 kg and $g = 9.81$ m s^{-2}.

a Show that Danny is travelling at 7.7 m s^{-1} at the top of the loop.

b Determine how much heavier Danny will feel when at the top of the loop compared to standing at rest.

Solution

a

Given	$v_{ramp} = 0.0$ m s^{-1}, $h = 7.0$ m, $r = 2.00$ m, $m_{D\ and\ bike} = 92$ kg, $g = 9.81$ m s^{-2}
Unknown	$v_{top\ of\ the\ loop} = ?$
Equations	$E_p = mg\Delta h$, $E_k = \dfrac{1}{2}mv^2$
Substitute	By the Law of Conservation of Energy:
	$E_{p\ ramp} = E_{p\ top\ of\ loop} + E_{K\ top\ of\ loop}$ and so
	$mg\Delta h = mg(2r) + \dfrac{1}{2}mv_{top\ of\ loop}^2$. Divide through by m and rearrange:
	$2(g\Delta h - 2gr) = v_{top\ of\ loop}^2$ so $v_{top\ of\ loop} = \sqrt{2g(\Delta h - 2r)}$
Solve	$v_{top\ of\ loop} = \sqrt{2 \times 9.81(7.0 - 2 \times 2.00)} = 7.67 = 7.7$ m s^{-1} (2 sf)

b The feeling of 'heaviness' is due to the reaction force pressing on Danny from the track and his bike pushing downwards on Danny at the top so he will feel heavy upwards.

Given	$v_{top\ of\ the\ loop} = 7.7$ m s^{-1}, $r = 2.00$ m, $m_D = 78$ kg, $g = 9.81$ m s^{-2}
Unknown	$F_{track\ top} = ?$ compared to $F_{track\ at\ rest} = ?$
Equations	$F_c = \dfrac{mv^2}{r}$, $F_g = mg$, $F_{c\ top} = F_{track} + F_g$
Substitute	$F_{track} = F_{c\ top} - F_g$ so $F_{track} = \dfrac{mv^2}{r} - mg$.
	Substituting in the numbers:
	$F_{track\ top} = \dfrac{78 \times 7.7^2}{2.00} - 78 \times 9.81$ so $F_{track} = 1547.1$ N
	When standing at rest, Danny has a reaction force on him of only
	$F_{track\ at\ rest} = 78 \times 9.81 = 765.2$ N
Solve	Danny will feel $1547.1 - 765.2 = $ **781.9 N heavier** at the top of the loop than when standing at rest. He will feel about twice as heavy but upwards.

Exercise 4E

Where required, take $g = 9.81 \text{ m s}^{-2}$.

1   Hannah is waiting for a bus with her sports bag full of gear. Initially, she holds the bag stationary, then she starts to swing the bag in a big arc.

   a   Draw labelled force vectors on the diagrams below to show all the forces acting on Hannah's hand as she walks. (A)

   b   Explain why the swinging bag feels heavier at the bottom of the swing compared to when she is just holding the bag. (M)

   _____

   _____

   _____

   _____

   _____

   The bag has a mass of 5.3 kg and the distance from her shoulder to the centre of mass of the bag is 1.2 m.

   c   Calculate the perceived weight of the bag as it moves at 0.95 m s^{-1} at the bottom of the swing.

   _____

   _____

   _____

2   During a traditional Maori cultural performance, Carol swings a poi around her hands at a steady speed. The head of the poi has a mass of 41.0 g and it is attached by a light string of length 30.0 cm.

   a   Show that the tension force in the string is 0.40 N when the poi is at rest and hanging down from her hand. (A)

   _____

   _____

   _____

b Show that an additional force of 1.21 N is required to make the poi move in a circle at a steady speed of 2.97 m s⁻¹. (A)

_____

_____

c Draw labelled force vectors to show the weight, tension force and net (centripetal) force for the positions shown in the diagram below. (A)

 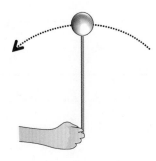

d Determine the tension force in the string when the poi is moving at 2.97 m s⁻¹ in the two positions shown above:

 i directly below her hand (A)

_____

_____

 ii directly above her hand. (A)

_____

_____

A photograph is taken of the poi when it is at an angle of 30° above the horizontal moving with a speed of 2.97 m s⁻¹.

e Draw labelled force vectors on the diagram to show the weight and tension force. (A)

f Draw the radial and tangential components of the weight vector on the diagram. (A)

g Calculate the size of the radial component of the weight that is acting towards the centre of the circle and use this to determine the tension force in the string. (M)

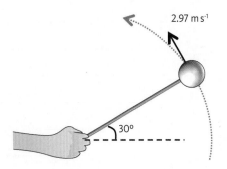

2.97 m s⁻¹

30°

_____

_____

_____

h Explain why Carol must actually move her hand in a small circle, about the centre of the circular path of the poi, to keep it moving at a steady speed. (E)

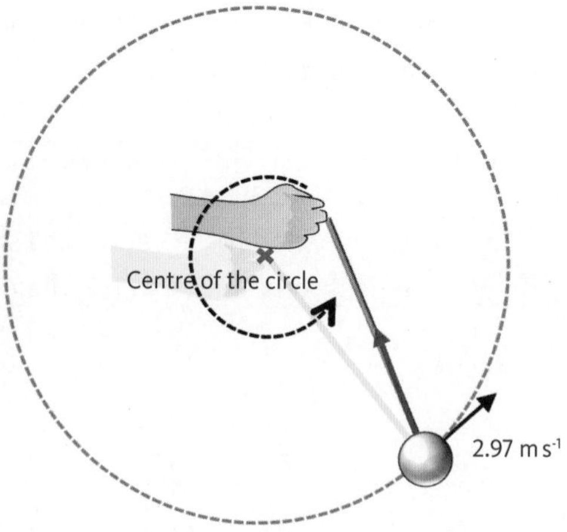

_____

_____

_____

_____

_____

_____

_____

Centre of the circle

2.97 m s⁻¹

To make the poi move at a constant speed, Carol must add energy to account for the increase in gravitational potential energy as the poi rises to the top. If Carol does not add this energy, the poi will slow down as it rises.

i Show that 2.41 J of gravitational energy is gained as the poi rises to the top.

_____

_____

j Determine the minimum amount of energy Carol should supply so that the poi just completes a vertical circle if it is moving at 2.97 m s⁻¹ at the bottom.

_____

_____

_____

_____

_____

_____

_____

3 A car full of passengers with a total mass of 1200 kg is driving along a straight road when the road drops down into a shallow dip and then rises back up over a bridge. The driver maintains a constant speed of 75.6 km h⁻¹ (21 m s⁻¹). The dip has a radius of arc of 45.0 m and the bridge has a radius of arc of 50.0 m.

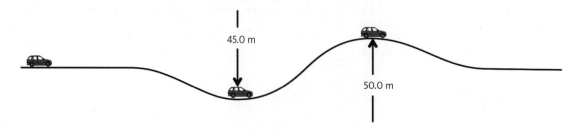

45.0 m

50.0 m

 ISBN: 9780170368179

a   Calculate the force from the road on the car as it passes through the bottom of the dip. (A)

_____

_____

_____

b   Explain how the passengers feel as they pass through the bottom of the dip. (M)

_____

_____

_____

_____

c   Calculate the force from the bridge on the car as it passes over the top of the bridge. (A)

_____

_____

_____

_____

d   Explain how the passengers feel as they pass over the top of the bridge. (M)

_____

_____

_____

_____

_____

e   Explain why taking the bridge at this speed is dangerous. (M)

_____

_____

_____

_____

_____

_____

f   Calculate the maximum speed at which the car could go over the bridge and just maintain contact with the road surface. Present your answer in km h⁻¹. (M)

_____

_____

_____

_____

_____

g Explain how the road engineers could change the dip and the bridge to enable cars to travel over this section at twice the current maximum speed, and discuss whether this change would be effective for all vehicles — large and small. (M)

_____

_____

_____

_____

_____

_____

_____

_____

_____

Gravitation
Newton's Law of Gravitation

In 1687, Newton published his greatest work, *Principia*, in which he proposed that the gravitational force that keeps planets in orbit around the Sun is exactly the same as the force that pulls an apple down to the surface of the Earth. Newton's hypothesis became known as the **Law of Universal Gravitation** and states:

> Every particle of matter, M, in the universe attracts every other particle of matter, m, with a force, F, that is proportional to their masses and inversely proportional to the square of the separation, r^2, of their centres of mass.

This can be presented as a formula:

$$F = \frac{GMm}{r^2}$$

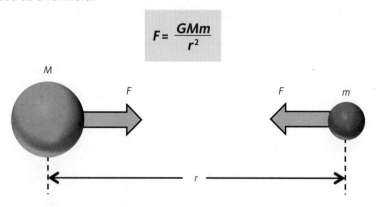

where the universal gravitational constant $G = 6.67 \times 10^{-11}$ N m^2 kg^{-2} is found by experiment. Due to the small size of G, the gravitational force of attraction between two objects is very small unless one of the objects is extremely massive.

Newton showed that this relationship was valid for:
- two spherically symmetrical bodies
- a spherically symmetrical body and an unsymmetrical body that is small compared with the separation of the masses
- two unsymmetrical bodies that are small compared with their separation.

Gravitational field strength, *g*

A gravitational field is a region around a massive object in which another mass experiences a force of attraction.

Near the surface of the Earth the ground appears flat and the gravitational field can be considered to be constant and vertically downwards. A 1 kg mass experiences a force of 9.81 N, which means the Earth's gravitational field strength **at the surface** is $g_{surface}$ = 9.81 N kg^{-1}, and if this mass is allowed to **fall freely** then it will accelerate at a rate of $a_{surface}$ = F/m = 9.81 m s^{-2}.

The gravitational field strength, *g*, at a specified point in space is given by the formula:

$$g = \frac{F}{m}$$

A uniform gravitation field acting downwards on the apple tree in Newton's garden.

which states that:

$$\text{gravitational field strength (N kg}^{-1}\text{)} = \frac{\text{force due to gravity (N)}}{\text{mass (kg)}}$$

However, moving away from the Earth reveals that it is not flat but spherical and the gravitational field acts radially towards the centre (similar in shape to that of an electric field around a negative charge).

As the distance from a mass increases, the field lines get further apart showing that the strength of the field decreases. So the gravitational field in space will be less than that at the surface of the Earth.

Combining Newton's Law of Gravitation and the gravitational field strength *g* gives the gravitational field strength at a distance *r* from an object of mass *M*:

$$g = \frac{GM}{r^2}$$

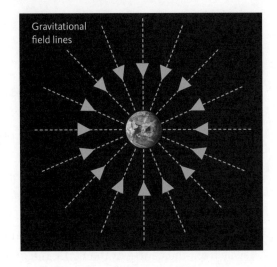

Gravitational field lines

which states that:

$$\text{gravitational field strength (N kg}^{-1}\text{)} = \frac{6.67 \times 10^{-11} \text{ (N m}^2 \text{ kg}^{-2}\text{) x mass of the object (kg)}}{\text{radius}^2 \text{ (m}^2\text{)}}$$

Worked example: Gravitational field strength and the International Space Station

The International Space Station maintains a low Earth orbit where the acceleration due to the Earth's gravity is 8.70 m s^{-2}. Calculate the height of the space station above the Earth's surface.

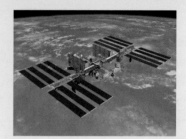

- Universal gravitational constant, $G = 6.67 \times 10^{-11}$ N m^2 kg^{-2}
- Mass of the Earth, $M_E = 5.972 \times 10^{24}$ kg
- Radius of the Earth, $R_E = 6.371 \times 10^6$ m

Solution

Given	$G = 6.67 \times 10^{-11}$ N m^2 kg^{-2}, $M_E = 5.972 \times 10^{24}$ kg, $R_E = 6.371 \times 10^6$ m, $a = 8.70$ m s^{-2}
Unknown	$h = ?$
Equations	$F = \dfrac{GMm}{r^2}$ and $F = ma$ (or $F = mg$). Combining gives $ma = \dfrac{GMm}{r^2}$ and rearranging:
Substitute	$R_{orbit} = \sqrt{\dfrac{GM}{a}} = \sqrt{\dfrac{(6.67 \times 10^{-11}) \times (5.972 \times 10^{24})}{8.70}} = 6.766 \times 10^6$ m
Solve	$h = R_{Orbit} - R_{Earth} = 6.766 \times 10^6 - 6.371 \times 10^6 = 395\ 000$ m above the Earth's surface

Extension concepts for scholarship — Variation of *g* with distance

The equation $g = \dfrac{GMm}{r^2}$ reveals that the gravitational field strength outside of a spherical object obeys an **inverse square law** (which is the case for all radial fields). So, for example, a person on the surface of the Earth experiences a gravitational field strength of $g_R = 9.81$ N kg^{-1}. If they then moved to a point three times further away from the centre of the Earth, the gravitational field strength *g* would reduce to ⅑ th of its original value, i.e. $g_{3R} = 1.09$ N kg^{-1}.

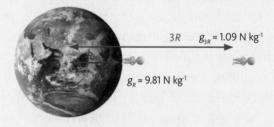

But what happens as you travel down inside the Earth? The distance *r* decreases as you dig down, but the amount of mass *M* below you decreases, and you start gaining mass above you!

Imagine the Earth was a hollow sphere with a person standing inside it. By dividing the sphere into two sections, it clearly shows that the green 'hemisphere' has less mass than the red 'hemisphere', but the green 'hemisphere' is closer, so the gravitational effect from the two sections is the same size but in the opposite directions, resulting in a gravitational field strength of $g_{hollow} = 0$ N kg^{-1} at every point inside a hollow sphere.

ISBN: 9780170368179

Now consider the Earth to be made of layers of hollow spheres (like an onion). As you dig down, each layer above you will have no effect on the gravitational field strength at that point. Consequently, the gravitational field strength inside the Earth only depends on the distance from the centre, r, and the remaining mass, M, as $g = GM/r^2$.

$g_r = 0 \text{ N kg}^{-1}$

If we assume that the density, ρ, of the Earth is uniform, $\rho = \frac{M}{V}$ so $M = \frac{4}{3}\pi r^3 \times \rho$ and:

$$g = \frac{4}{3} G\pi r \rho$$

This reveals that inside a solid sphere of uniform density, the gravitational field strength **increases linearly** with distance from the centre. All these ideas can be simple expressed in a graph of the variation in g with distance from the centre of mass.

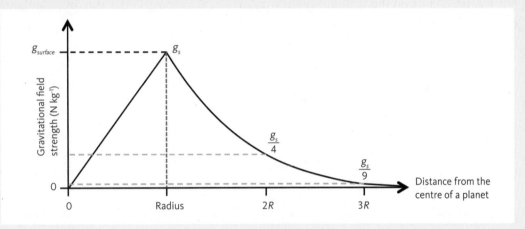

Exercise 4F

Unless otherwise stated, take:
- Universal gravitational constant, $G = 6.67 \times 10^{-11}$ N m^2 kg^{-2}
- Earth's gravitational field strength, $g = 9.81$ N kg^{-1}
- Mass of the Earth, $M_E = 5.972 \times 10^{24}$ kg
- Radius of the Earth, $R_E = 6.371 \times 10^6$ m
- Mass of the Moon, $M_M = 7.35 \times 10^{22}$ kg
- Orbital distance of the Moon, $r_M = 384 \times 10^6$ m.

1   Kate and William are standing with their centres of mass 80.0 cm apart. Kate has a mass of 58 kg and William has a mass of 72 kg.
    a   Calculate the gravitational force of attraction between them.

    b   Determine the mass of an object that would have the same weight force on Earth. Give your answer in milligrams.

2   Objects on Earth feel heavy due to the effects of the Earth's gravitational field.

   a   Calculate the force due to gravity on a 1.00 kg mass at the surface of the Earth using the formula $F_g = \frac{GMm}{r^2}$. (A)

   _____

   _____

   b   Calculate the force due to gravity on the same mass when placed at a distance of 6.766×10^6 m, the same distance from the centre of the Earth as the International Space Station. (A)

   _____

   _____

   The force on the 1.00 kg mass is only 11% less on the International Space Station than on Earth, but the astronauts on the station would describe a 1.00 kg mass in the station as 'weightless'.

   c   Explain the apparent contradiction. (E)

   _____

   _____

   _____

   _____

   _____

   _____

   _____

   _____

   _____

   _____

   _____

3   Apollo 15 Commander David Scott dropped a 1.32 kg hammer onto the lunar surface. The hammer experienced a force of 2.145 N pulling it downwards. The Moon has a diameter of 3.474×10^6 m.

   a   Show that the mass of the Moon is 7.35×10^{22} kg. (M)

   _____

   _____

   b   Calculate the gravitational field strength at the surface of the Moon. (M)

   _____

   _____

   The force due to the gravitational fields around the Moon and the Sun are able to affect the tides on Earth.

   c   Show that the Moon's gravitational field exerts a force of 34.4 µN on each kilogram of water at the Earth's surface. (A)

   _____

   _____

d Calculate the distance of the centre of the Sun from the surface of the Earth given that the Sun has a mass of 1.989×10^{30} kg and exerts a force of 59.28×10^{-4} N on each kilogram of water on the surface. (A)

_____

_____

e Explain how the gravitational force from the Sun on the water is so much greater than the force from the Moon even though it is about 400 times further away. (M)

_____

_____

_____

_____

_____

_____

_____

4 The Moon's gravitational field strength at the surface of the Moon is about $\frac{1}{6}$ th of the Earth's gravitational field strength at that surface of the Earth.
   a Explain this difference in the gravitational field strength between the two bodies. (E)

_____

_____

_____

_____

_____

_____

_____

The centres of the Earth and Moon are 384 000 km apart but their gravitational fields overlap in space. The Lagrange point is a region between them where the combined gravitational field strength will be zero.
   b Determine the distance from the centre of the Moon to the Lagrange point. (E)

_____

_____

_____

_____

_____

_____

_____

Scholarship questions

5 Anneke and Eddie are discussing an experiment carried out by Robert Hooke in the late 1600s. He attempted to prove that the force due to gravity at the Earth's surface was stronger than at the top of a tall building. He did this by measuring a 1.0 kg mass at the bottom and top of Westminster Abbey, a height of 60 m. Anneke and Eddie decide to replicate the experiment by taking a newton meter, with an accuracy to ±1 Nm, up 222 m to the top floor of the Sky Tower.

Compare and contrast the two experiments and discuss the validity of the results.

ISBN: 9780170368179

6   Study the following graphs of gravitational field strength by distance from the centre of an object, discuss what they tell you about the structure of the objects and decide which one most accurately models the Earth.

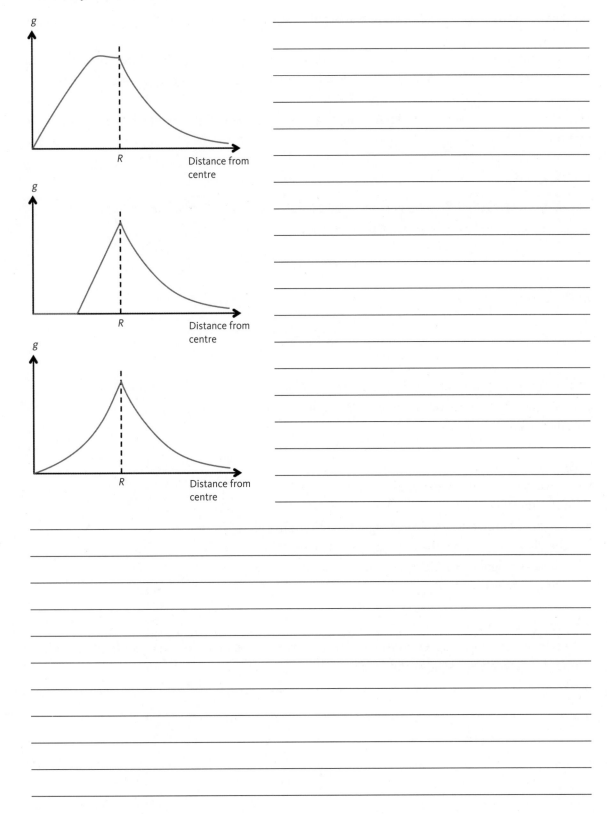

_____

_____

_____

_____

_____

_____

_____

_____

_____

_____

_____

_____

_____

_____

_____

_____

_____

_____

_____

_____

_____

_____

_____

_____

_____

_____

_____

_____

_____

Satellite motion

Any natural object that orbits a larger mass due to its gravitational field can be described as a satellite.

- Natural satellites — the Earth and planets are natural satellites of the Sun, and the Moon is a natural satellite of the Earth.
- Artificial satellites — any manmade objects that are put into orbit around the Earth, Moon, Sun or other planets, for example communication satellites, GPS satellites, research satellites.

Many orbits are circular or very slightly elliptical, so can be modelled effectively using circular motion.

Stable orbits

A stable orbit occurs when the gravitational force provides the necessary centripetal force required for the satellite to move in a circle, hence:

$$\frac{GMm}{r^2} = \frac{mv^2}{r}$$

which rearranges to give:

$$v = \sqrt{\frac{GM}{r}}$$

This represents the orbital speed, v, of a satellite at an orbital radius, r, from the centre of mass of a larger primary object of mass, M.

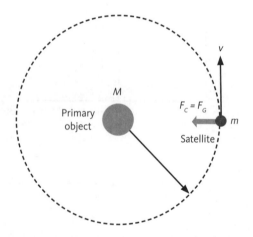

Note:

i   The orbital speed is independent of the mass of the orbiting object.

ii   Beware of problems involving height (or altitude)! 'r' is the distance between the centres of the two objects, whereas height is the distance from the surface of the primary object, so $r = R_{primary} + h$.

Period of an orbit

For a satellite orbiting at speed v around a circular path of circumference $C = 2\pi r$, it takes one time period, T, to complete a single orbit, where:

$$T = \frac{d}{v} = \frac{2\pi r}{\sqrt{\dfrac{GM}{r}}}$$

so

$$T = 2\pi \sqrt{\frac{r^3}{GM}}$$

Polar orbits

A satellite in a low Earth orbit over the equator will travel very quickly with a time period of 90 minutes. This enables it to pass over the whole surface of the Earth as the Earth slowly rotates underneath it during the course of a day. This is ideal for earth-monitoring satellites such as for weather.

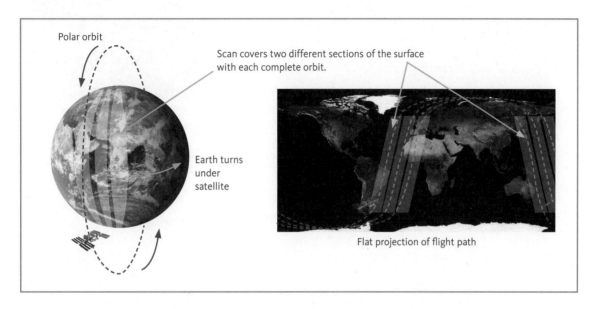

Geostationary (geosynchronous) orbits

A satellite in a high Earth orbit over the equator travels slower. If its time period matches the rotation of the Earth (i.e. 24 hours), it will remain permanently above a single point on the Earth. This is ideal for communication satellites, as their location remains fixed in the sky.

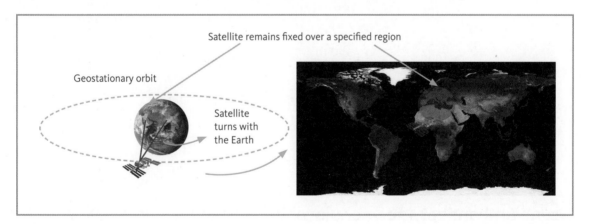

Worked example: International Space Station revisited

The International Space Station maintains a low Earth orbit at a height of 395 km above the surface of the Earth. Calculate the time taken to complete a single orbit of the Earth. Give your answer in hours.

- Universal gravitational constant, $G = 6.67 \times 10^{-11}$ N m² kg⁻²
- Mass of the Earth, $M_E = 5.972 \times 10^{24}$ kg
- Radius of the Earth, $R_E = 6.371 \times 10^6$ m

Solution

Watch out: the question provides height (in km), but the formulae rely on radial distance (in m) from the centre!

Given	$h = 395 \times 10^3$ m, $R_E = 6.371 \times 10^6$ m, $G = 6.67 \times 10^{-11}$ N m² kg⁻², $M_E = 5.972 \times 10^{24}$ kg
Unknown	$T = ?$
Equations	$F_g = \dfrac{GMm}{r^2}$, $F_c = \dfrac{mv^2}{r}$, $v = \dfrac{d}{t}$, $C = 2\pi r$, $r = R_E + h$
Substitute	Radius of the orbit: $r = R_E + h = (6.371 \times 10^6) + (395 \times 10^3) = 6.766 \times 10^6$ m In a stable orbit: $F_c = F_g$, so $\dfrac{mv^2}{r} = \dfrac{GMm}{r^2}$, which simplifies to give $v = \sqrt{\dfrac{GM}{r}} = \sqrt{\dfrac{GM}{R_E + h}}$ so $v = \sqrt{\dfrac{(6.67 \times 10^{-11}) \times (5.972 \times 10^{24})}{6.766 \times 10^6}} = 7672.85$ m s⁻¹. The time period can now be found using: $T = \dfrac{2\pi r}{v} = \dfrac{2\pi \times 6.766 \times 10^6}{7672.85} = 5540.58$ s
Solve	Time to complete an orbit: $T = \dfrac{5540.58}{3600} = 1.54$ hours

Extension concepts for scholarship — Spin versus gravity

Planets and stars are generally spherical in nature, but if they are spinning rapidly they start to bulge at the equator and become oblate (like a squashed sphere). Photographs reveal Saturn's oblate shape due to its extremely fast rotation speed.

If a planet or star rotates at a very fast speed, then the gravitational force acting on matter at the surface may not be strong enough to provide the centripetal force necessary to keep the matter in contact with the surface (in the same way that a car skids off a road if the friction force is too small). As a consequence, the planet or star would disintegrate.

Consider an object at the equator of a non-rotating planet. The gravitational force between the planet of mass M and the object of mass m is given by the formula:

$$F_g = \frac{GMm}{r^2}$$

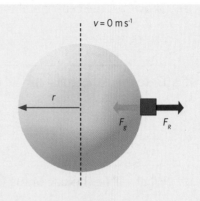

On a stationary planet, the gravitational force is pulling the object down onto the surface resulting in a reaction force F_R. The perceived weight of an object is due to the reaction force and so as:

$$F_R = F_g$$

the object feels as heavy as expected.

Now if the planet is rapidly spinning, a component of the gravitational force provides a centripetal force maintaining the object's motion in a circle. The remaining component of the gravitational force pulls the object onto the surface so:

$$F_R = F_g - F_c$$

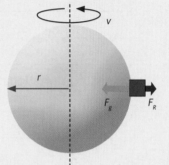

The reaction force is now less, so the perceived weight of the object is less than the weight measured on the non-rotating planet. From the equation it can be seen that at:
- slow rotational speed, $F_g > F_c$ and the object will exert a reduced force on the ground
- faster rotational speed, $F_g = F_c$ and the object will not exert any force on the ground, so $F_R = 0$ N and the object will orbit and appear weightless
- very fast rotational speed, $F_g < F_c$, matter on the surface will lose contact with the surface, and the planet disintegrates.

There is no centripetal force required at the poles, so the reaction force will be equal to the gravitational force.

Exercise 4G

Unless otherwise stated, take:
- Universal gravitational constant, $G = 6.67 \times 10^{-11}$ N m^2 kg^{-2}
- Earth's gravitational field strength, $g = 9.81$ N kg^{-1}
- Mass of the Earth, $M_E = 5.972 \times 10^{24}$ kg
- Radius of the Earth, $R_E = 6.371 \times 10^6$ m
- Mass of the Moon, $M_M = 7.35 \times 10^{22}$ kg
- Orbital distance of the Moon, $r_M = 384 \times 10^6$ m.

1   A satellite of mass 1450 kg is put into orbit at a distance of 8240 km from the centre of the Earth. It takes 124.1 minutes to complete each orbit.

   a   Show that the satellite is travelling with a speed of 6.95 km s^{-1}. (A)

_____

_____

b Show that the centripetal force on the satellite 8.51 kN. (A)

_____

_____

c Determine the gravitational force on the satellite and explain your answer. (M)

_____

_____

_____

d What will be the size of the force of the satellite on the Earth? Explain your answer. (A)

_____

_____

_____

e Calculate the gravitational field strength at this distance from the Earth. (A)

_____

_____

2 Mars has two moons — Phobos and Deimos — which are thought to be captured asteroids. Phobos orbits Mars at a speed of 2138 m s^{-1} with an orbital radius of 9376 km. It is held in orbit by the gravitational force between Mars and Phobos of 5.191×10^{15} N.

a Show that the time period of Phobos is 7 hours 39 minutes 14 seconds. (A)

_____

_____

b Calculate the mass of Phobos. (A)

_____

_____

_____

c Determine the mass of Mars. (M)

_____

_____

3 NOAA-15 is a global weather-monitoring satellite that was placed in a low Earth polar orbit in 1998. It has a mass of 1457 kg and travels once around the Earth every 101.1 minutes at an altitude of 813 km above the Earth's surface.

a Show that the satellite has an orbital radius of 7184 km. (A)

_____

_____

b Calculate the orbital speed. (A)

_____

_____

c State the conditions necessary for a stable orbit. (A)

_____

_____

_____

_____

d Determine the mass of the Earth. (E)

_____

_____

_____

4 GOES-13 is a satellite with a mass of 3133 kg that is used to monitor solar flare activity. It is in a high Earth orbit with a radius of 42.23 x 106 m.

    a Show that the orbital velocity of GOES-13 is 3.07×10^3 m s^{-1}. (M)

_____

_____

_____

_____

_____

_____

b Prove that GOES-13 is a geostationary satellite and explain your working. (M)

_____

_____

_____

_____

GOES-13's orbital radius is about six times greater than the orbital radius of a polar orbiting satellite like NOAA-15 but the orbital period is over 14 times longer.

    c Explain why the increase in the time period is significantly greater than the increase in the orbital radius. (E)

_____

_____

_____

_____

_____

GOES-13 shares its orbit with millions of tiny pieces of rock and dust that have a mass less than a few grams.

d Explain how it is possible for objects with different mass to move in the same orbit. (E)

_____

_____

_____

_____

_____

_____

_____

_____

e Determine the height of GOES-13 above the surface of the Earth. (A)

_____

_____

Scholarship question

5 During the formation of a planet, the gravitational force is responsible for holding matter together as the planet cools and solidifies. However, if the forming planet is rotating too quickly, matter can be lost if the gravitational force is not strong enough.

For a planet with a uniform density ρ and volume $V = \frac{4}{3}\pi r^3$, show that the minimum period of rotation is given by the formula:

$$T = \sqrt{\frac{3\pi}{G\rho}}$$

_____

_____

_____

_____

_____

_____

_____

_____

_____

_____

_____

_____

_____

_____

4.2 Rotational motion

Adopting rotational thinking

Rotational motion can be quickly understood by simply changing three aspects of the approach to translational motion:

- Change from '**Up and right are positive**' to '**Anticlockwise is positive**' approach.
- Change the measurement of motion from **metres (translational)** to **radians (rotational)**. See Chapter 1 Angles and circles, page 6, for an explanation of radians.
- Change concepts from using linear quantities to rotational quantities, e.g. **forces (translational)** → **torques (rotational)**. See below for further details.

These three simple changes in approach mean that rotational problems can be solved using the same skills and ideas developed in the study of translational motion at Levels 1 and 2.

Part 1: Rotational motion

Translational distance to angular displacement, θ (theta)

Consider two particles A and B attached to a rod that is slowly rotating. Particle B has a greater radius and as a consequence travels a greater **distance**, *d*, in the same time as A. However, both A and B travel through the same **angular displacement**, θ, such that:

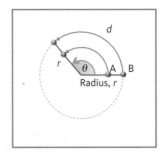

$$d = r\theta$$

which states that:

linear distance (m) = radius (m) x angular displacement (rad)

Translational velocity to angular velocity, ω (omega)

Translational velocity, *v*	Angular velocity, ω
A particle travelling at a **steady speed** *v* along a circular path will cover a **distance** *d* in a time *t*, hence: $$v = \frac{\Delta d}{\Delta t}$$ which states that: velocity $(m\ s^{-1}) = \frac{\text{change in displacement (m)}}{\text{change in time (s)}}$	A particle travelling at a **steady angular velocity** ω along a circular path of constant radius r will **cover an angular displacement** θ in a time *t*, hence: $$\omega = \frac{\Delta\theta}{\Delta t}$$ which states that: angular velocity $(rad\ s^{-1}) = \frac{\text{change in angular displacement (rad)}}{\text{change in time (s)}}$

As $d = r\theta$, so for a circular path of constant radius, $v = \dfrac{r\Delta\theta}{\Delta t}$, and as $\omega = \dfrac{\Delta\theta}{\Delta t}$:

$$v = r\omega$$

which states that: **linear speed (m s⁻¹) = radius (m) x angular velocity (rad s⁻¹)**

For particle B to complete the circle in the same time as particle A, it must travel at a greater **velocity** as it covers a greater **distance**. However, both A and B sweep through the same **angular displacement**, θ, in the same time, so travel at the same **angular velocity**, ω.

Translational acceleration to angular acceleration, α (alpha)

Translational acceleration, a	Angular acceleration, α
The **acceleration**, a, of a particle travelling along a circular path will result in a change in **tangential speed**, Δv, in a time t, hence: $$a = \frac{\Delta v}{\Delta t}$$ which states that: $$\text{acceleration (m s}^{-2}) = \frac{\text{change in velocity (m s}^{-1})}{\text{change in time (s)}}$$	The **angular accleration**, α, of a particle along a circular path of constant radius r will result in a change in **angular velocity**, ω, in a time t, hence: $$\alpha = \frac{\Delta \omega}{\Delta t}$$ which states that: $$\text{angular acceleration (rad s}^{-2}) = \frac{\text{change in angular velocity (rad s}^{-1})}{\text{change in time (s)}}$$

As $v = r\omega$, so for a circular path of constant radius, $a = \dfrac{r\Delta\omega}{\Delta t}$, and as $\alpha = \dfrac{\Delta\omega}{\Delta t}$:

$$a = r\alpha$$

which states that:

$$\boxed{\text{linear acceleration (m s}^{-2}) = \text{radius (m) x angular acceleration (rad s}^{-2})}$$

Periodic motion, frequency, angular velocity and revolutions

Rotational motion is described as periodic motion, as it regularly repeats the same cycle as the object spins around.

The time period, T, is the time taken for a spinning object to complete one cycle,

an angular displacement of 2π radians. As $\omega = \dfrac{\Delta\theta}{\Delta t}$, so: $\quad \omega = \dfrac{2\pi}{T}$

The number of rotations each second is described by the frequency of the rotation: $\quad f = \dfrac{1}{T}$

Hence the angular velocity can be expressed in terms of the frequency: $\quad \omega = 2\pi f$

The total number of revolutions, N, completed by a rotating object can be found by dividing the

angular displacement, θ (rad), by 2π: $\quad N = \dfrac{\theta}{2\pi}$

Rotational equations of motion

For constant angular acceleration situations in which the change in velocity is uniform and takes place along a circular path of constant radius, these basic equations can be combined to produce five equations of motion, which can accurately describe the movement of an object.

Translational	Relationship	Rotational
$d = \left(\dfrac{v_f + v_i}{2}\right) t$	Equation of motion 1	$\theta = \left(\dfrac{\omega_f + \omega_i}{2}\right) t$
$v_f = v_i + at$	Equation of motion 2	$\omega_f = \omega_i + \alpha t$
$d = v_i t + \dfrac{1}{2} at^2$	Equation of motion 3	$\theta = \omega_i t + \dfrac{1}{2} \alpha t^2$
$v_f^2 = v_i^2 + 2ad$	Equation of motion 4	$\omega_f^2 = \omega_i^2 + 2\alpha\theta$
$d = v_f t + \dfrac{1}{2} at^2$	Equation of motion 5	$\theta = \omega_f t + \dfrac{1}{2} \alpha t^2$

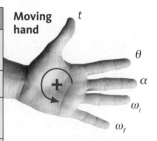

Moving hand

Time is assigned to the **T**humb because it is the **only** scalar quantity. Each finger represents a vector quantity.

When solving equations of motion problems, use the **Moving hand** and **GUESS** techniques. Each term can be assigned to a digit on your hand to remind you to check which terms have been **G**iven by the question, and which term is the **U**nknown.

Each equation of motion contains FOUR terms and has one term missing. Ticking off each term on your hand makes choosing the correct **E**quation from the formula list very straightforward. The values can then be **S**ubstituted and **S**olved.

Be careful: some quantities can be hidden in the wording. For example:

- an object may be described as being 'stationary' or 'at rest'; both these statements mean it has zero angular velocity
- an object travelling at a steady angular velocity will have a zero angular acceleration.

Worked example: Compact disc player

When a compact disc player is started, the CD accelerates from rest up to 5.00×10^2 rpm in 3.0 seconds. The laser starts reading the inner track of the disc and gradually moves outwards. The inner track has a diameter of 46 mm.

a Calculate the angular acceleration of the disc during the first three seconds.

b Show that the translational speed of the inside track is 1.2 m s^{-1} when spinning at its operating speed.

c The outer track has a diameter of 110 mm and has to spin at the same translational speed as the inside track to allow the laser to read the information.

Calculate the number of complete revolutions the CD completes as the laser moves from the inside track to the outside track while playing music for 72 minutes.

Solution

Initially, convert the given information to SI units, so $f_{final} = 500$ rpm $= \dfrac{500}{60} = 8.33$ Hz, and $d_{track} = 46$ mm $= 46 \times 10^{-3}$ m.

a

Given	$\omega_i = 0$ rad s^{-1}, $f_{final} = 8.33$ Hz, $t = 3.0$ s, $d = 46 \times 10^{-3}$ m
Unknown	$\alpha = ?$
Equations	$\omega_f = \omega_i + \alpha t$ and $\omega = 2\pi f$
Substitute	$\omega_f = 2\pi \times 8.33 = 52.36$ rad s^{-1} hence $52.36 = 0 + \alpha \times 3.0$. Rearranging:
Solve	$\alpha = \dfrac{52.36 - 0}{3.0} = 17.45 = 17$ rad s^{-1} (2 sf)

b

Given	f_{final} = 8.33 Hz, d = 46 x 10^{-3} m, and, from above, ω = 52.36 rad s^{-1}
Unknown	v = ?
Equations	$v = r\omega$ and $r = \dfrac{d}{2}$ = 23 x 10^{-3} m
Substitute	v = 23 x 10^{-3} x 52.36 hence
Solve	v = 1.2 m s^{-1} (2 sf)

c

Given	v_f = 1.2 m s^{-1}, ω_i = 52.36 rad s^{-1}, d_{outer} = 110 x 10^{-3} m
Unknown	θ = ?, and the number of revolutions.
Equations	$v_f = r_{outer}\,\omega_f$, $r_{outer} = \dfrac{d}{2}$ = 55 x 10^{-3} m, $\theta = \left(\dfrac{\omega_f + \omega_i}{2}\right) t$, and no. of revs = $\dfrac{\theta}{2\pi}$
Substitute	1.2 = 55 x 10^{-3} x ω_f hence ω_f = 21.818 rad s^{-1}
	so $\theta = \left(\dfrac{21.818 + 52.362}{2}\right)$ x 72 x 60 = 160224.87 rad
Solve	No. of revs = $\dfrac{\theta}{2\pi} = \dfrac{160224.87}{2\pi}$ = 25500 = 26000 revolutions (2 sf)

Exercise 4H

1 The wheel of a racing bicycle has a radius of 0.311 m and Sarah rolls the bicycle forwards 2.00 m at a steady speed.

    a Determine the angular displacement of the front wheel. (A)

    b Calculate the angular velocity if it takes him 1.2 s to move this distance. (A)

The cycle race starts and Sarah accelerates from rest to a speed of 11 m s^{-1} in 8.0 s.

    c Calculate the linear acceleration of the bicycle. (A)

    d Calculate the angular acceleration of the wheel. (A)

2 The Sky Tower restaurant Orbit rotates through a complete revolution every hour, offering patrons the chance to see all of Auckland and the Hauraki Gulf. The restaurant rests on a large ring with an outer diameter of 16.0 m.

    a Show that the angular velocity of the Sky Tower is 1.7 x 10^{-3} rad s^{-1}. (A)

When the power is switched on in the morning, the restaurant turns through 2.6 x 10^{-2} rad before reaching full speed.

    b Calculate the acceleration of the restaurant. (A)

c Determine the time taken for the restaurant to reach full speed. (A)

A large party is dining in the restaurant one night. One diner is sitting by the window at the outer edge of the ring. A diner sits at the other end of the table near the tower a distance of 3.0 m from the windows.

d Determine the difference in the distance travelled by someone sitting by the window and someone sitting near the tower if they both spend 2.4 hours eating their meal. (M)

3 A toy airplane is attached to the top of a pole by a 2.00 m-long rod that allows it to fly in horizontal circles. When released, the motor causes the airplane to accelerate at 0.77 rad s⁻² from rest up to its top speed in 5.2 seconds.

2.00 m

a Show that the angular displacement of the airplane is 10.4 rad by the time it reaches its top speed. (A)

b Determine the number of revolutions the airplane completes by the time it reaches its top speed. (A)

c Calculate the final angular velocity of the airplane. (A)

Sometime later the battery on the airplane starts to run down. The airplane slows from 3.00 rad s⁻¹ to 0.60 rad s⁻¹ while completing 43 revolutions.

d Calculate the angular acceleration of the airplane during this time. (M)

e Calculate the average tangential speed of the airplane while it is slowing down. (A)

4 The main rotor blades on a helicopter have a radius of 3.4 m and the tip of the blade travels at a tangential speed of 192 m s⁻¹ while hovering.

a Calculate the frequency of revolution of the helicopter blades in rpm. (M)

b Compare and contrast the tangential speed and angular velocity of a point on a rotor blade halfway along blade. (M)

_____

_____

_____

After the helicopter lands, the rotor blades slow down from 50.0 rad s^{-1} to 35.0 rad s^{-1} and sweep out an angle of 7650 rad.

c Calculate the time taken to change the angular velocity. (A)

_____

_____

d Calculate the angular acceleration of the rotor blades while slowing down.

_____

The pilot is left waiting on the tarmac as the blades continue to slow down, and they are still moving when a call comes through requiring an immediate take-off. The rotor blades accelerate at 0.30 rad s^{-2} to an angular velocity of 58.0 rad s^{-1} in 2.4 minutes.

e Calculate the number of revolutions the blades complete while accelerating. (E)

_____

_____

_____

_____

_____

Part 2: Rotational forces and momentum

Not only can you compare translational and rotational motion, you can also compare forces, mass, the principles of momentum and Newton's laws of motion.

Translational force to angular force — torques, τ (tau)

For a **force**, **F**, to cause an object to start rotating or change its rate of rotation (angular velocity), it must act at a distance r from a pivot or the centre of mass. The size of the **torque**, τ, is given by the formula:

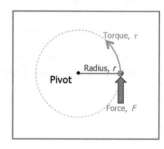

$$\tau = Fr$$

which states that:

torque (N m) = force (N) x radius (m)

Translational mass (inertia) to angular mass — rotational inertia, I

The concept of **translational mass (inertia)**, m, relates the difficulty of changing a body's **translational motion** due to its **mass**. Similarly, the rotational inertia, I, relates the difficulty of changing a body's rotation about an axis. However, for a rotating object, the distance of the mass from the axis also affects the rotational inertia. (Try lifting a bag close to your body, then at arm's length. The same mass 'feels heavier' when lifted at arm's length — that's rotational inertia.)
The rotational inertia, I, of a point mass is given by the formula:

$$I = mr^2$$

which states that:

rotational inertia (kg m²) = mass (kg) x radius² (m²)

The rotational inertia depends upon an object's mass, shape, and the position of the axis about which the object is being made to rotate.

The rotational inertia of more complex objects can be described by the general formula:

$$I = \Sigma mr^2$$

which states that:

rotational inertia of a complex object = the sum of (Σ) all the individual rotational inertias

and considers a complex object to be made up of lots of individual point masses at different distances. Students are not expected to perform this operation, but they must be aware of the general relationship. Some sample rotational inertias are given below:

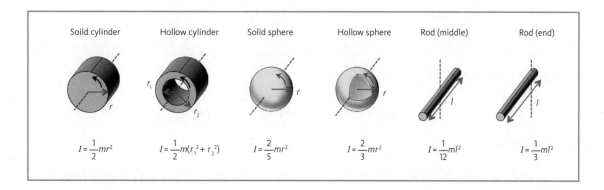

Soild cylinder	Hollow cylinder	Solid sphere	Hollow sphere	Rod (middle)	Rod (end)
$I = \frac{1}{2}mr^2$	$I = \frac{1}{2}m(r_1^2 + r_2^2)$	$I = \frac{2}{5}mr^2$	$I = \frac{2}{3}mr^2$	$I = \frac{1}{12}ml^2$	$I = \frac{1}{3}ml^2$

Translational momentum to angular momentum, L

Translational momentum, p	Angular momentum, L
An object with **mass m** travelling at a **steady speed v** along a straight path will have a **momentum p** of: $$p = mv$$ which states that: momentum = mass x velocity (kg m s⁻¹) (kg) (m s⁻¹)	An object with **rotational inertia I** travelling at steady **angular velocity ω** along a circular path will have an **angular momentum L** of: $$L = I\omega$$ which states that: angular momentum = rotational inertia x angular velocity (kg m² s⁻¹) (kg m²) (rad s⁻¹)
The principle of **conservation of translational momentum** states that: The total **translational momentum** of a system will remain constant provided that there is no net external **force** acting on the system. This can be expressed mathematically as: **total p_i = total p_f**	The principle of **conservation of angular momentum** states that: The total **angular momentum** of a system will remain constant provided that there is no net external **torque** acting on the system. This can be expressed mathematically as: **total L_i = total L_f**
Applying a **force, F**, to the object will cause a change in **momentum, p**, as $F = \dfrac{\Delta p}{\Delta t}$, and so for constant mass m, we have: $F = \dfrac{m\Delta v}{\Delta t}$. As $a = \dfrac{\Delta v}{\Delta t}$, this simplifies to: $$F_{net} = ma$$ Newton's second law (translational)	Applying a **torque, τ**, to the object will cause a change in **angular momentum, L**, as $\tau = \dfrac{\Delta L}{\Delta t}$, and so for constant rotational inertia I, we have: $\tau = \dfrac{I\Delta \omega}{\Delta t}$. As $a = \dfrac{\Delta \omega}{\Delta t}$, this simplifies to: $$\tau_{net} = I\alpha$$ Newton's second law (rotational)

Substituting $I = mr^2$ and $\omega = \dfrac{v}{r}$ into $L = I\omega$ gives:

$$L = mvr \quad \text{and as } p = mv, \text{ so} \quad L = pr$$

which states that:

angular momentum (kg m² s⁻¹) = translational momentum (kg m s⁻¹) x radius (m)

This formula is useful when a rotating object is released and travels away at a tangent, e.g. a cricket bowler releasing a ball, or when an object moving in a straight line is captured and starts to rotate, e.g. a child running and jumping on a swing.

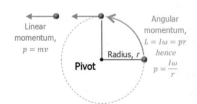

Linear momentum, $p = mv$

Angular momentum, $L = I\omega = pr$ hence $p = \dfrac{I\omega}{r}$

Radius, r

Pivot

   ISBN: 9780170368179

Newton's laws of rotational motion

All of Newton's translational laws have an equivalent rotational form as shown below.

	Linear	Rotational
Newton's FIRST law	An object will remain at rest or continue moving at a **constant speed** in the same direction unless acted on by a net external force.	An object will remain at rest or continue moving at a constant **angular velocity** in the same direction unless acted on by a net external **torque**.
Newton's SECOND law	The rate of change of **translational momentum** of an object is directly proportional to the net external **force** acting upon it, and takes place in the direction of the net force. OR for constant **mass**: The **translational acceleration** of an object is directly proportional to the net external **force** acting upon it. $$F = ma$$	The rate of change of **angular momentum** of an object is directly proportional to the net external torque acting upon it, and takes place in the direction of the net **torque**. OR for constant **rotational inertia**: The **angular acceleration** of an object is directly proportional to the net external **torque** acting upon it. $$\tau = I\alpha$$
Newton's THIRD law	If object A exerts a **force** on object B, then object B will exert an equal but oppositely directed **force** on object A.	If object A exerts a **torque** on object B, then object B will exert an equal but oppositely directed **torque** on object A.

 ## Worked example: The toy helicopter

A bucket full of water, mass 10.2 kg, is suspended over a well by a light rope wrapped around a frictionless pulley of radius 20.0 cm. The system is initially held at rest by a brake, which is then released causing the bucket to accelerate into the well at 2.30 m s^{-2}. Take the acceleration of gravity as 9.81 m s^{-2}.

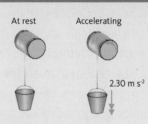

a   Draw force vector arrows on the diagrams to show the weight and tension forces acting on the bucket.

b   Explain why the weight of the bucket does not cause it to accelerate at 9.81 m s^{-2}.

c   Calculate the tension force acting on the bucket when it is accelerating.

d   Determine the rotational inertia of the pulley.

Solution

a   For the system to remain at rest, the forces acting on the stationary bucket must be balanced. The accelerating bucket must experience a net force downwards, so the tension force on the bucket must be smaller than the weight force.

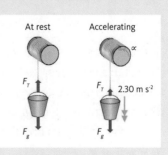

b   The weight force must accelerate the rotational inertia of the pulley as well as the inertial mass of the bucket, so the acceleration of the system is less than if the force were just acting on the mass alone.

c

Given	$m = 10.2$ kg, $a = 2.30$ m s^{-2}, $g = 9.81$ m s^{-2}
Unknown	$F_T = ?$
Equations	$F_{net} = ma$, $F_g = mg$ and $F_{net} = F_g - F_T$
Substitute	$F_{net} = ma = 10.2 \times 2.30 = 23.46$ N and $F_g = mg = 10.2 \times 9.81 = 100.062$ N
Solve	Tension force on the bucket: $F_T = F_g - F_{net} = 100.062 - 23.46 = 76.602 = 76.6$ N (3 sf)

d

Given	$a = 2.30$ m s^{-2}, $r = 0.200$ m, $g = 9.81$ m s^{-2}
Unknown	$I = ?$
Equations	$\tau = Fr$, $a = r\alpha$ and $\tau = I\alpha$
Substitute	$\tau = Fr = 76.6 \times 0.200 = 15.32$ Nm and $\alpha = \dfrac{a}{r} = \dfrac{2.30}{0.200} = 11.5$ rad s^{-2}
Solve	Rotational inertia: $I = \dfrac{\tau}{\alpha} = \dfrac{15.32}{11.5} = 1.332 = 1.33$ kg m^2

Extension concepts for scholarship — Rotational vectors and gyroscopes

Rotational quantities are vectors so must have magnitude and direction. As the vector direction of the tangential motion at the edge of a spinning object is always changing, we use the right-hand grip rule to define the vector direction of rotational quantities. The fingers of the right hand point in an anticlockwise direction leaving the thumb indicating the positive vector direction of rotational quantities.

Positive vector direction

Spin direction

Gyroscopes demonstrate the amazing properties of this vector nature of rotational quantities, and are constructed from a rapidly spinning mass on a low-friction bearing. They are used in airplane autopilot systems, Segways and self-balancing electric scooters, but the basic principles apply to much more everyday phenomena, like why bicycles are very stable when they are moving.

Consider a person standing at rest on a freely rotating platform; the initial angular momentum:

$$L_{i\,p+p} = 0 \text{ kg m}^2 \text{ s}^{-1}$$

They are holding a rapidly spinning bicycle wheel (gyroscope) as shown in the photograph:

$$L_{i\,gyro} = L \uparrow$$

If the person then applies a torque to the gyroscope, they can turn it over so that it is spinning in the opposite direction. But this causes the person to start spinning on the platform in the original direction of the wheel. How?

Consider the final angular momentum of the gyroscope:

$$L_{f\,gyro} = L \downarrow$$

 ISBN: 9780170368179

The torque provided by the person is an internal torque, so momentum is conserved.

Which means:

$$L_{i\,system} = L_{f\,system}$$

hence:

$$L_{i\,gyro} + L_{i\,p+p} = L_{f\,gyro} + L_{f\,p+p}$$

Substituting in values:

$$L\uparrow + 0 = L\downarrow + L_{f\,p+p}$$

hence:

$$L\uparrow - L\downarrow = L_{f\,p+p}$$

Swapping the sign and direction:

$$L\uparrow + L\uparrow = L_{f\,p+p} \text{ so } L_{f\,p+p} = 2L\uparrow$$

By the principle of conservation of angular momentum, the person and platform will start to spin in the original direction of the wheel.

Exercise 4I

1   A toy helicopter works by pulling on a string that is wrapped around the axle of the helicopter rotor. The axle has a diameter of 3.00 cm. The rotor has a mass of 25.0 g and a diameter of 10.0 cm and its rotational inertia can be modelled on a thin hollow cylinder using the formula $I = mr^2$.

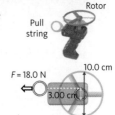

Rotor
Pull string
$F = 18.0$ N
10.0 cm
3.00 cm

   a   Show that the rotational inertia of the rotor is 6.25×10^{-5} kg m^2. (A)

   b   Determine the angular acceleration of the rotor when a force of 18.0 N is applied by pulling the string. (M)

   c   Explain what would happen to the angular acceleration if a rotor with three times the rotor radius and double the mass is used instead. The axle radius and force remain the same.

2   A load is suspended by a long string wrapped around a pulley of radius 16.0 cm and mass 8.60 kg. The pulley can be modelled on a solid cylinder with a rotational inertia of $I = \frac{1}{2}mr^2$. When released, the mass accelerates downwards at 5.274 m s^{-2}. Take $g = 9.81$ m s^{-2}.

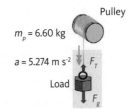

Pulley
$m_p = 6.60$ kg
$a = 5.274$ m s^{-2} F_T
Load
F_g

   a   Show that the rotational inertia of the wheel is 0.110 kg m^2. (A)

   b   Show that the torque which accelerates the fly wheel is 3.63 Nm. (M)

c   Calculate the mass of the load. In your solution you should include:
   i   a labelled force vector diagram to show the forces acting on the load
   ii an explanation of how you know that these forces are unbalanced
   iii the tension force in the string that is producing the torque on the pulley. (E)

_____

_____

_____

_____

_____

_____

3   In the 1950s, gyrobuses were developed that contained a flywheel with a rotational inertia of 88 kg m^2. At each bus stop an external power supply was used to accelerate the flywheel up to a speed of 3361 rpm. The flywheel has a radius of 0.80 m and can be modelled on a solid cylinder, $I = \frac{1}{2} mr^2$.

a   Calculate the mass of the flywheel. (A)

_____

b   Explain how the flywheel could be redesigned to use half the mass but still have the same rotational inertia. (M)

_____

_____

_____

c   Calculate the average **net** torque required to accelerate the flywheel from rest up to its top speed in 1.5 minutes and explain why the torque applied to the flywheel would need to be greater. (E)

_____

_____

_____

_____

_____

One of the reasons that the gyrobuses went out of use was the tendency of the metal in the flywheel to suffer from fatigue and fail catastrophically sending bits of metal flying off at high speed.

When spinning at its top speed a 0.500 kg piece of the flywheel comes loose from the edge of the flywheel and travels vertically upwards.

d   Show that the angular momentum of the flywheel is 31 x 10^3 kg m^2 s^{-1} before the piece flies off. (A)

_____

_____

e   Explain why angular momentum is conserved when the piece breaks loose and hence describe the motion of the flywheel and the piece immediately after it breaks free. In your answer you should consider both momentum and speed of each part. No calculations are required. (E)

_____

_____

_____

_____

_____

_____

f   Show that the tangential velocity of the 0.500 kg piece of the flywheel is 280 m s^{-1} at the instant it leaves the flywheel. (A)

_____

_____

g   Calculate the rotational inertia of the flywheel after the piece flies off.

_____

_____

_____

4   An astronaut on the ISS is testing the relationship between angular velocity and radius. He rotates a stone of mass 0.15 kg on the end of a piece of string above his head in a horizontal circle of radius 0.90 m. The time taken to complete a single rotation is 0.50 s. The rotational inertia of the stone can be determined using $I = mr^2$. 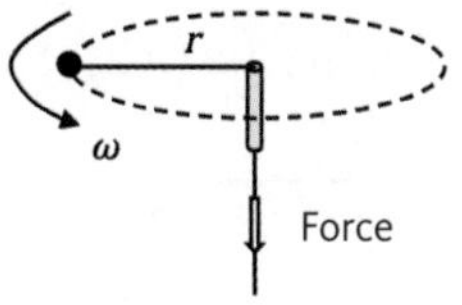
You may assume friction is negligible.

a   Determine the initial angular momentum of the stone. (M)

_____

_____

_____

The string passes through a smooth tube allowing it to be pulled in so that the radius is reduced to 0.30 m.

b   Explain why angular momentum will be conserved even though an external force acts on the string. (A)

_____

_____

_____

c   Determine the final angular velocity of the stone. (M)

_____

_____

_____

d Determine how much faster the stone is moving after the radius decreases, and compare it to the radius. (A)

_____

_____

5 Amelia, Ellis, Daniel and Jake are at the park playing on a roundabout, which spins on a frictionless axle in the centre of the wheel. It has a mass of 175 kg and a rotational inertia 283.5 kg m². Amelia has a mass of 40.0 kg and stands 0.50 m away from the centre of the roundabout. Ellis has a mass of 30.0 kg and stands 1.00 m from the centre of the roundabout. Jake has a mass of 20.0 kg and stands 1.50 m from the centre of the roundabout.

a By modelling the roundabout on a thin solid cylinder with a rotational inertia of $I = \frac{1}{2} mr^2$, show that the radius of the roundabout is 1.80 m. (A)

_____

_____

b By modelling each of the children as a point mass with a rotational inertia of $I = mr^2$, show that the total rotational inertia of the roundabout and children combined is 369 kg m². (M)

_____

_____

_____

_____

c Explain why Jake has a larger rotational inertia than Amelia despite having the smaller _inertia_ (mass). (E)

_____

_____

_____

_____

_____

_____

The roundabout is initially stationary, so to get it spinning Daniel runs at a tangent to the edge and jumps onto the roundabout 1.70 m from the centre, as shown in the diagram. Daniel has a mass of 48.0 kg and is initially moving at 5.0 m s⁻¹ when he jumps onto the roundabout and grabs the rail.

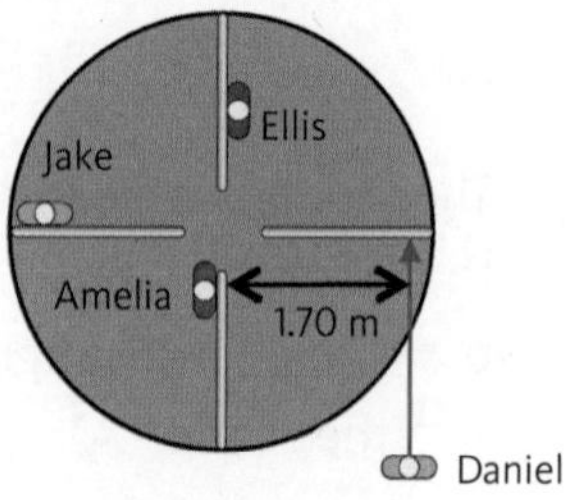

d Show that the angular velocity of Daniel just before he lands on the roundabout is 410 kg m² s⁻¹. (A)

_____

e Determine the angular velocity of the system (roundabout and children) after Daniel lands on the roundabout. Justify your solution with relevant physics principles. (E)

_____

_____

_____

_____

_____

_____

_____

_____

_____

Daniel bends his arms as he hits the rail and so it takes him 0.80 s to come to a stop relative to the roundabout.

f   Calculate the force that is applied to Daniel by the rail. (E)

_____

_____

_____

_____

_____

_____

_____

While the roundabout is still spinning, all the children walk to the centre.

g   Discuss what will happen to the angular momentum of the roundabout. (E)

_____

_____

_____

_____

_____

At the end of the ride the children get off feeling a bit dizzy. Jake puts his arms out to try to keep his balance.

h   Explain why humans put their arms out to help them balance. (M)

_____

_____

_____

_____

Scholarship question

6 A bicycle, resting on the ground, is being lightly held in an upright position by its lower pedal. The pedals are vertical and a horizontal backwards force, F_p, is applied to the lower pedal, a distance r_p from the front axle. The chain transfers the pedal force to the back wheel of radius r_w.

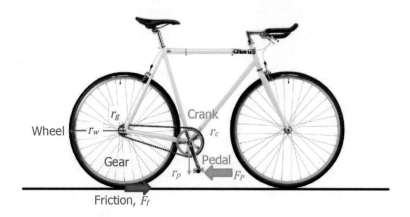

Prove that:

$$F_f = F_p \frac{r_p r_g}{r_w r_c}$$

and hence discuss in which direction the bicycle, the wheel and the pedal will move.

_____

_____

_____

_____

_____

_____

_____

_____

_____

_____

_____

_____

_____

_____

_____

_____

_____

_____

Part 3: Rotational energy

Translational (linear) kinetic energy and rotational kinetic energy, $E_{K\,rot}$

Translational (linear) kinetic energy, E_K	Rotational kinetic energy, $E_{K\,rot}$
An object with a **mass m sliding** sideways with a **translational speed v** has **translational (linear) kinetic energy $E_{K\,lin}$**: $$E_{K\,lin} = \frac{1}{2}mv^2$$ which states that: kinetic energy $= \frac{1}{2}$ x mass x velocity^2 (J) (kg) (m^2 s^{-2}) 	An object with a **rotational inertia I spinning** with a **rotational speed ω** has **rotational kinetic energy $E_{K\,rot}$**: $$E_{K\,rot} = \frac{1}{2}I\omega^2$$ which states that: rotational kinetic energy (J) $= \frac{1}{2}$ x rotational inertia (kg m^2) x angular velocity^2 (m^2 s^{-2})

A rolling object **moves sideways** and **spins** so it will have both translational and rotational kinetic energy.

$$E_{KT} = E_{K\,lin} + E_{K\,rot} \quad \text{so} \quad E_T = \frac{1}{2}mv^2 + \frac{1}{2}I\omega^2$$

Note: **translational velocity** and **tangential velocity** are the same for a rolling object and $v = r\omega$.

Translational work and rotational work, W

Work must be done to bring about a change in the energy of an object.

Translational work, W	Rotational work, W
Work is done, W, on an object when a **force** F moves the object through a **translational distance d** in the direction of the **force**, hence: $$W = Fd$$ which states that: work (Nm) = force (N) x distance (d)	**Work is done**, W, on an object when a **torque** τ moves the object through an **angular displacement** θ in the direction of the **torque**, hence: $$W = \tau\theta$$ which states that: work (Nm) = torque (Nm) x angular displacement (rad)

ISBN: 9780170368179

Principle of Conservation of Energy

> Energy cannot be created or destroyed, only transformed from one type to another.

Consider a block and a rolling cylinder accelerating down a slope. The gravitational force does work on both objects causing a change in energy.

If friction effects are negligible, then a sliding object transforms all its $E_{P\,grav}$ into **translational** $E_{K\,lin}$ as it accelerates down the slope. The block will be moving fast when it reaches the bottom, as $E_{K\,lin} = \frac{1}{2}mv^2$.

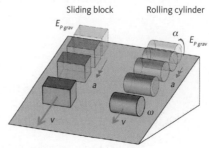

Sliding block   Rolling cylinder

For the sliding block
$E_{K\,lin} = E_{P\,grav}$

For the rolling cylinder
$E_{K\,lin} = E_{P\,grav} - E_{K\,rot}$

A rolling object transforms some of the energy into **rotational** $E_{K\,rot}$ as well as **translational** $E_{K\,lin}$. This means that when the rolling cylinder reaches the bottom, it will be moving slower than the block, as it has less **translational** $E_{K\,lin}$.

 ## Worked example: Sisyphus's boulder solution

Having made several frustrating attempts trying to push a spherical boulder to the top of a hill, only to watch it roll back down again, Sisyphus decides on a new, more cunning plan. He decides to get a run-up with the boulder, then let it go at the base of the hill. If it is moving fast enough, it will gradually slow down and come to rest just as it reaches the top of the hill. The hill is 48.0 m high, the boulder has a mass of 2.00×10^2 kg, a radius of 1.80 m and can be modelled on a sphere with a rotational inertia of $I = \frac{2}{5}mr^2$. Take $g = 9.81$ N kg^{-1}.

Determine the minimum speed at which the boulder must be moving at the base of the hill for Sisyphus's plan to succeed and discuss whether the task would be easier with a sliding boulder.

Solution

Given	$h = 48.0$ m, $m = 2.00 \times 10^2$ kg, $r = 1.80$ m, and $v_f = 0$ m s^{-1}, $\omega = 0$ rad s^{-1}
Unknown	$v_i = ?$
Equations	$E_P = mgh$, $E_{K\,lin} = \frac{1}{2}mv^2$, $E_{K\,rot} = \frac{1}{2}I^2$, $I = \frac{2}{5}mr^2$ and $v = r\omega$

Substitute   The principle of conservation of energy states that the kinetic energy at the base of the hill must be equal to the gravitational potential energy at the top of the hill, assuming that no energy is lost, for example, to heat due to friction doing work on the boulder.

Therefore: $E_{K\,lin\,base} + E_{K\,rot\,base} = E_{P\,top}$ so $\qquad \frac{1}{2}mv_i^2 + \frac{1}{2}I\omega^2 = mgh$

Substituting in for I and ω gives: $\qquad \frac{1}{2}mv_i^2 + \frac{1}{2}\left(\frac{2}{5}mr^2\right)\left(\frac{v_i}{r}\right)^2 = mgh$

Simplifying: $\qquad \frac{1}{2}mv_i^2 + \frac{1}{5}mv_i^2 = mgh$

Dividing through by m and combining terms: $\qquad \frac{7}{10}v_i^2 = gh$

Rearranging: $\qquad v_i = \sqrt{\dfrac{10gh}{7}}$

Solve   Substituting in values: $v_i = \pm 25.9$ m s^{-1}

This is nearly 100 km h^{-1} so Sisyphus's plan is extremely unlikely to succeed, as energy lost to heat and sound will require the boulder to be moving even faster to reach the top. The minimum speed of a sliding boulder will need to be even faster, as it only has linear kinetic energy at the base, so $E_{K\,lin\,base} = E_{p\,top}$ hence $v_i = \sqrt{2gh} = 30.7$ m s^{-1} and frictional forces will be much more significant with a sliding object. Sisyphus should try being an author as an alternative to pushing boulders uphill!

Note: The mass has no effect on the solution but does increase the friction force between the boulder and the hillside.

Exercise 4J

1   A cylinder of mass 0.750 kg, radius 6.00 cm and rotational inertia 1.35×10^{-3} kg m^2 is released from rest at the top of a smooth slope. It rolls down the slope until it reaches the end with a tangential speed of 3.00 m s^{-1}. Take $g = 9.81$ m s^{-2}.

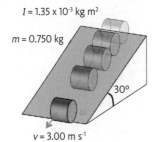

$I = 1.35 \times 10^{-3}$ kg m^2

$m = 0.750$ kg

$v = 3.00$ m s^{-1}

a   Calculate the change in translational kinetic energy. (A)

_____

_____

b   Calculate the change in rotational kinetic energy. (M)

_____

_____

c   Hence show that the slope is 0.688 m high. (M)

_____

_____

d   If the slope has an angle of 30° to the horizontal, calculate the length of the slope. (M)

_____

_____

_____

The cylinder is then placed on its end so that it can slide down the same smooth slope.

e   Determine the difference in translational speed between the rolling cylinder and the sliding cylinder. Explain your solution and state any assumptions that you have made. (M)

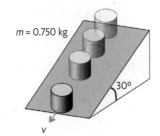

$m = 0.750$ kg

v

_____

_____

_____

_____

2   In 2013, Volvo demonstrated a flywheel system in one of its sedan cars. When the car brake pedal is pressed the flywheel is connected to the road wheels causing energy from the road wheels to go into spinning the flywheel up to 6.00×10^4 rpm instead of being transferred to heat through conventional brakes. The flywheel is made of carbon fibre with a mass of 6.00 kg and it has a diameter of 0.20 m.

   a   Calculate the angular velocity of the flywheel. (A)

   _____

   b   Calculate the rotational inertia of the flywheel if it is modelled on a solid cylinder, $I = \frac{1}{2}mr^2$.

   _____

   c   Determine the maximum energy that could be stored in the flywheel.

   _____

   An alternative flywheel design is based on a hollow cylinder with a rotational inertia given by $I = mr^2$ but both flywheels have the same mass and radius.

   d   Discuss how changing to the new flywheel will affect the operation of the flywheel. In your answer you should consider:
   - the maximum energy stored by the flywheel
   - the effect of the same force on the acceleration of the flywheel
   - the time taken to reach the same rotational frequency of 6.00×10^4 rpm.

   _____
   _____
   _____
   _____
   _____
   _____
   _____
   _____
   _____
   _____
   _____

3   Jane is a trampoline gymnast who is practising jumps and spins. Each time she jumps, some of the energy of the trampoline goes into providing height and some goes into making her spin. Jane is 1.68 m tall, has a mass of 56.0 kg and can jump, without spinning, to a maximum height of 10.00 m above the trampoline.

   a   Show that the maximum energy supplied by Jane and the trampoline is 5490 J. (A)

   _____

   The first jump Jane makes involves a single spin while keeping her body straight.

   b   Calculate Jane's rotational inertia by modelling her body on a rod spinning about the centre, $I = \frac{1}{12}ml^2$. (A)

   _____
   _____

c   Calculate the maximum height Jane can reach if she can just complete a single rotation during the jump with an angular velocity of 2.2 rad s^{-1}. (E)

_____

_____

_____

_____

_____

_____

_____

Jane can reduce her rotational inertia to ¼ of the original value by changing her body shape from being straight to being in a tucked position.

d   Explain how going into a tuck halfway through the jump changes her rotational motion. In your answer you should consider:
   •   why going into a tuck changes her rotational inertia
   •   what happens to her angular momentum
   •   what happens to her angular velocity as a result
   •   what happens to her rotational kinetic energy.

_____

_____

_____

_____

_____

_____

_____

_____

_____

_____

_____

_____

_____

_____

_____

_____

_____

Scholarship questions

4 The seat of a stool is made from a screw-threaded column that allows users to increase its height by spinning the seat anticlockwise. The seat has a mass m and radius r and can be modelled on a solid cylinder with a rotational inertia of $I = \frac{1}{2}mr^2$.

Determine the height the seat rises when given an initial tangential velocity v. You may assume that there is no friction between the seat column and the frame.

Explain whether angular momentum is conserved.

_____

_____

_____

_____

_____

_____

_____

_____

_____

5 Alastair is designing a ski jump, so he builds a model and rolls a marble of radius r, mass m and rotational inertia $I = \frac{2}{5}mr^2$ down the track to look at how it performs. The marble is released from rest at a height h above the lowest point in the track. The track is shown in the diagram with the jump having a radius of R.

mass, m

h

R

a Show that the maximum height of release of the ball so that it remains in contact with the track is given by the relationship $h = \frac{17}{10}R$.

_____

_____

_____

_____

_____

_____

_____

_____

 ISBN: 9780170368179

4.3 Oscillatory motion

Oscillatory motion is a form of periodic motion and describes a continuously repeating action in which a particle is displaced, either side of an equilibrium position.

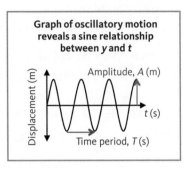

Graph of oscillatory motion reveals a sine relationship between y and t

Examples of oscillatory motion include: a swinging pendulum, a mass bouncing on a spring, a piston in a car engine, a vibrating guitar string, an electron inside an alternating current, a boat bobbing on water, the rise and fall of tides, a ball rolling across a round bowl.

Oscillatory motion is described in terms of three fundamental properties.

- **Displacement**, y (m) — the distance from the equilibrium position.
- **Amplitude**, A (m) — the maximum displacement, y, from the equilibrium position.
- **Time period**, T (s) — the time taken to complete a single oscillation.
- **Frequency**, f (Hz) — the number of oscillations each second.

$$f = \frac{1}{T}$$

Simple harmonic motion, SHM

SHM definition

Simple harmonic motion (SHM) is one type of oscillatory motion and occurs when:

> The acceleration, a, of an object acts towards the equilibrium position and is proportional to the displacement, y, away from it.

This is expressed mathematically as shown below and can be used to identify SHM.

$$a = -\text{(constant)}y$$

Note: The constant is positive and dependent on the type of SHM.

The − sign indicates that the acceleration is in the opposite direction to the displacement.

SHM and the principle of conservation of energy

During SHM energy is constantly being transformed from potential to kinetic and back again, however the total amount of energy in the system remains constant unless friction does work changing energy to heat, sound, etc. The diagrams show the energy changes for a ball rolling back and forth across the bottom of a round bowl, but the concept applies to all SHM oscillators.

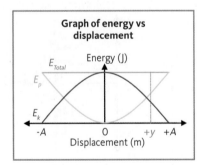

Graph of energy vs displacement

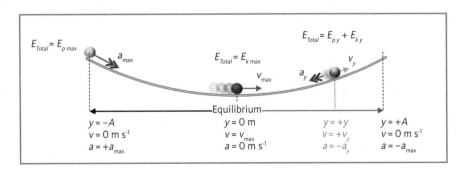

Damping

In reality SHM oscillators gradually lose energy due to effects such as friction, drag, etc. Over time, damping causes the total energy available to decrease, which means that:

- maximum potential energy decreases — amplitude decreases
- maximum kinetic energy decreases — maximum speed decreases.

However, the frequency of the oscillation remains constant, unless critical damping is applied causing the energy to dissipate rapidly to zero without oscillating (see the graphs below).

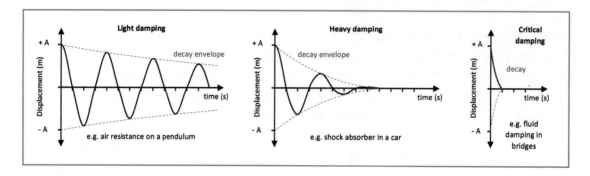

Resonance

All objects have a **natural or resonant frequency** (f_o) of oscillation that occurs when the body is displaced from its equilibrium position and then released, e.g. plucking a ruler on the edge of a desk. If a body is subjected to a periodic external force (driving force), then it will start to oscillate in **forced SHM** (f_D).

Resonance occurs when the frequency of the driving force is equal to the natural frequency of the object causing the amplitude of the oscillations to increase as energy keeps being added to the system at a greater rate than it is lost.

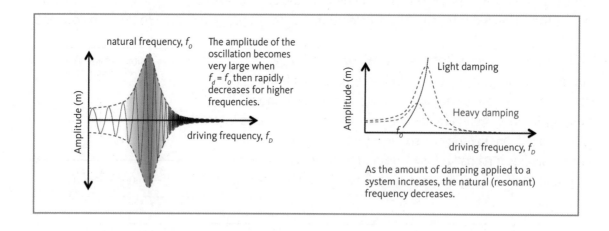

   ISBN: 9780170368179

Worked example: The shocking truth

Henry is out driving in his car when he goes over a bump in the road. This causes the car to bounce up and down five times before it finally stops oscillating. The maximum amplitude of the oscillations is 3.5 cm and the time period of the oscillations is 0.6 s.

Draw a graph to show how the amplitude decreases with time due to the shock absorbers.

Solution

1   The graph needs to include 5 oscillations so mark the x-axis into 10 divisions (2 per wave) every 0.3 s.

2   Mark the maximum amplitudes above and below the equilibrium.

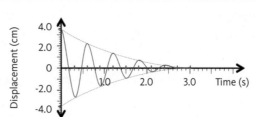

3   Draw in a heavy damping wave envelope symmetrically either side of the equilibrium — this is easiest to do if you turn the page around so your hand is inside the curve. The envelope should finish at the end of the fifth oscillation.

4   Unless specifically stated in the question, you can start the wave at either amplitude or at the equilibrium. The initial force from the bump causes the shock absorber to go upwards so the oscillation starts from the top amplitude.

5   Sketch in pencil a smooth sine curve with a time period 0.6 s. Check your graph and then go over it in pen.

Examples of SHM oscillators

Mass oscillating on a spring

Consider a mass, m, attached to a spring with a spring constant, k, at rest on a frictionless surface. If the spring is compressed a distance, y, and then released the mass will experience a net force: $\qquad F_{net} = -ky$

causing it to accelerate to the left: $\qquad F_{net} = ma$

After the spring reaches its normal length it will start to exert a force to the right slowing the mass down until it eventually stops and starts travelling in the opposite direction. The mass will continue to oscillate horizontally.

The graph to the right shows how the acceleration varies with displacement and leads to the equation of the line:

$$a = -\left(\frac{k}{m}\right)y$$

which is of the same form as the equation for SHM showing that the mass oscillates in simple harmonic motion.

Time period, T

The time period, T, of a mass oscillating on a spring is given by the formula:

$$T = 2\pi \sqrt{\frac{m}{k}}$$

Energy

Elastic potential energy stored in a stretched or compressed spring:

$$E_P = \frac{1}{2} ky^2$$

Kinetic energy of moving mass:

$$E_K = \frac{1}{2} mv^2$$

Total energy

At some point y in the oscillation:

$$E_{Total} = E_{Py} + E_{Ky} = \frac{1}{2} ky^2 + \frac{1}{2} mv^2$$

At maximum displacement $v = 0$ m s^{-1} so:

$$E_{Total} = E_{P\,max} + 0 = \frac{1}{2} kA^2$$

At equilibrium, $y = 0$ m so:

$$E_{Total} = 0 + E_{K\,max} = \frac{1}{2} mv_{max}^2$$

The simple pendulum and SHM

When a mass, m, suspended from a pivot by a light string of length l is pulled to one side and released it will oscillate back and forth tracing out a small arc. At either amplitude (maximum displacement) the bob is briefly at rest so not accelerating centripetally as:

$$F_{Tension} = F_{gll}$$

But the perpendicular component of weight:

$$F_{g\perp} = mg \sin \varphi$$

produces a net force on the bob:

$$F_{net} = ma$$

which acts towards the equilibrium position and is proportional to the displacement causing the bob to accelerate:

$$a = g \sin \varphi$$

as $d = r\varphi$. For small angles ($\varphi < 10°$) then $\sin \varphi \approx \varphi$ (in radians) and so:

$$\varphi = \frac{y}{l}$$

which gives:

$$a = -\left(\frac{g}{l}\right) y$$

(The negative sign is included to show that the acceleration acts in the opposite direction to the displacement.)

This equation is of the same form as the equation for SHM showing that the pendulum oscillates in simple harmonic motion when the angle is small.

Time period, T

The time period, T, of a simple pendulum is given by the formula:

$$T = 2\pi \sqrt{\frac{l}{g}}$$

   ISBN: 9780170368179

Energy

Gravitational potential energy stored in the mass when raised:

$$E_P = mgh$$

 where $h = l - s$ and $s = l \cos \varphi$ so $h = l - l \cos \varphi$

Kinetic energy of moving mass:

$$E_K = \frac{1}{2}mv^2$$

Total energy

At some point y in the oscillation:

$$E_{Total} = E_{Py} + E_{Ky} = mgh + \frac{1}{2}mv^2$$

At maximum displacement $v = 0$ m s^{-1} so:

$$E_{Total} = E_{P\,max} + 0 = mgA$$

At equilibrium, $y = 0$ m so:

$$E_{Total} = 0 + E_{K\,max} = \frac{1}{2}mv_{max}^2$$

Worked example: Rope swings

Ed is playing on a swing. Each oscillation takes 3.50 s and he travels a total distance of 2.00 m back and forth. Determine

a   the amplitude of his oscillation
b   the length of the chain. State any assumptions that you make.
Use $g = 9.81$ m s^{-2}.

Solution

a   The amplitude of an oscillation is from the equilibrium position to the maximum displacement. During any single oscillation Ed passes through 4 amplitudes,

   so $A = \dfrac{2.00}{4} = 0.50$ m.

b

Given	$T = 3.50$ s and $g = 9.81$ m s^{-2}
Unknown	$l = ?$
Equations	$T = 2\pi\sqrt{\dfrac{l}{g}}$
Substitute	Squaring both sides, $T^2 = 4\pi^2 \times \dfrac{l}{g}$. Multiplying both sides by $\dfrac{g}{4\pi^2}$ gives $T^2 \times \dfrac{g}{4\pi^2} = l$
Solve	$l = 3.50^2 \times \dfrac{9.81}{4\pi^2} = 3.04$ m (3 sf)

This assumes that:

1   the chain ends at Ed's centre of mass. As the weight force drives the pendulum, the distance from the pivot must always be measured to the centre of mass where the force acts.
2   the angle of the swing is less than 10° so that the small angle approximation applies. (Using the amplitude and chain length we have $\theta = \dfrac{d}{r} = \dfrac{0.50}{3.04} = 0.164$ rad $= \dfrac{0.164}{2\pi} \times 360 = 9.42°$.)

Exercise 4K

1   A mass is attached to a spring causing it to extend a distance $x = 13.08$ cm
    before coming to rest. The spring constant $k = 30.0$ N m^{-1}.

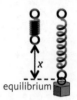

   a   Show that the restoring force $F_{Rx} = -3.92$ N using Hooke's law $F_{Rx} = -kx$
       and explain what the negative sign in the statement indicates. (A)

   _____

   _____

   _____

   b   State the size of the weight force and explain how you determined your answer. Include an
       algebraic equation to show the net force, F_{net}. (A)

   _____

   _____

   _____

   The mass is then pulled down an additional distance, y, and released
   causing it to accelerate upwards.

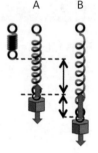

   c   Draw labelled force vectors on the diagram opposite showing
       the forces acting on the mass when in equilibrium (A) and the
       moment after release (B).

   d   By considering the forces on the mass in equilibrium and the net
       force the moment after it is released, show that the maximum
       acceleration is given by the equation:

   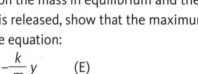

$$a = -\frac{k}{m}y \qquad (E)$$

   _____

   _____

   _____

   _____

   _____

   _____

   e   Using the equation above, explain how you can tell that the mass is moving in simple
       harmonic motion. (A)

   _____

   _____

   _____

   _____

A new unknown mass is hung from the spring and displaced from the equilibrium by 2.00 cm before being released. A graph of acceleration versus displacement for the new mass is shown opposite.

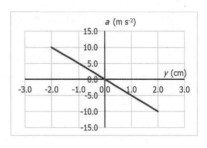

f Calculate the gradient of the graph, then using the equation $a = -\dfrac{k}{m} y$ determine the new mass hanging on the spring. (M)

_____

_____

_____

2 Joel has a mass of 14 kg and is oscillating up and down in a baby bouncer — a large spring with a harness at the bottom. Joel is suspended so he just can't reach the ground and then pulled down slightly so that he begins to bounce. Each complete bounce takes 0.80 s and he travels a distance of 12.0 cm from the bottom to the top.

a Determine the frequency of the oscillations. (A)

_____

_____

_____

b Show that the spring constant of the spring is 860 N m⁻¹. (M)

_____

_____

_____

Energy is stored in the spring when Joel is just sitting at rest in the equilibrium position, but to start it bouncing requires additional energy. (E)

c Determine the amplitude of the oscillation.

_____

_____

d At what point in the oscillation will Joel be moving at his fastest speed?

_____

_____

e Determine Joel's maximum speed by considering the additional energy stored in the spring when Joel is at the bottom of a bounce.

_____

_____

_____

_____

Joel's maximum speed when bouncing gradually decreases until he comes to a stop after about 10 seconds.

f   Draw a graph of displacement against time to show how his amplitude changes over 4 oscillations. Explain why this change occurs. (E)

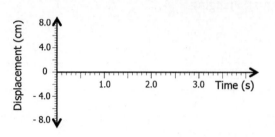

_____

_____

_____

_____

_____

_____

_____

_____

_____

_____

g   Joel's acceleration is given by the formula $a = -\dfrac{k}{m}\, y$. By considering this relationship, sketch a graph below showing how his acceleration changes over the same period of time. Justify your answer. (E)

_____

_____

_____

_____

_____

_____

_____

_____

_____

The baby bouncer has a weight restriction on it for safety reasons.

h   Explain how the amplitude, time period and frequency would change if a child with twice the mass tried to bounce in the baby bouncer. (E)

_____

_____

_____

_____

_____

_____

_____

3 The door to the glove box in Simon's car has an annoying rattle which only occurs when the tachometer (rev counter) in his car tells him that the engine is spinning at 1300–1450 rpm. This is annoying as it means the car rattles a lot when driving at a steady speed of 50 km h^{-1}.

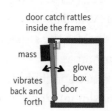

door catch rattles inside the frame

mass

vibrates back and forth

glove box door

Simon thinks that the door could be modelled on an upside-down simple pendulum and attempts to fix the rattle by attaching a mass near the top of the door.

a Discuss Simon's rattle problem and solution. (E)

In your answer you should:

- determine the average frequency at which the door of the glove box rattles
- explain the relationship between engine speed and the rattle
- comment on the success of his solution and suggest an alternative solution.

_____

_____

_____

_____

_____

_____

_____

_____

_____

_____

_____

_____

_____

_____

_____

Simon drives to a restaurant where he notices that the door to the kitchen swings both in and out to allow serving staff to rapidly get to and from the kitchen. After a member of staff passes through the door it swings back and forth three times before coming to rest. The door is released a distance of 0.70 m from the frame and swings with a frequency of 0.40 Hz.

b Draw a graph to show how the displacement varies with time until the door comes to rest. Include values where possible.

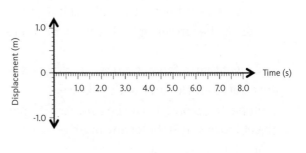

Reference circles — comparing SHM to circular motion

SHM and circular motion appear to be very different but they are both forms of periodic motion. Consider the situation opposite with a pendulum swinging in SHM and a mass going round on a turntable.

The pendulum starts swinging from its maximum amplitude, accelerates towards the equilibrium position, then slows down to a stop at the other maximum amplitude. The mass on the turntable goes around at a steady speed covering regular distances in regular times. However, if we watch the shadows of the pendulum and the mass cast by the lamp, they move together in perfect unison.

This provides us with a method of describing mathematically the position, velocity and acceleration of an object moving with SHM using the motion of an imaginary particle around a reference circle.

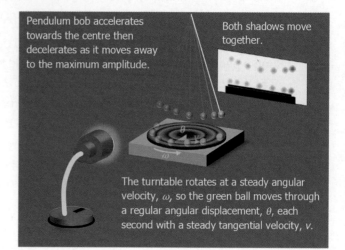

Pendulum bob accelerates towards the centre then decelerates as it moves away to the maximum amplitude.

Both shadows move together.

The turntable rotates at a steady angular velocity, ω, so the green ball moves through a regular angular displacement, θ, each second with a steady tangential velocity, v.

Reference circles

The motion of the object moving in SHM can be described in terms of the motion of the imaginary particle around the reference circle:

- **amplitude, A (m)**
  The maximum displacement of the object moving in SHM. It is the same as the radius of the reference circle.

  $$y_{max} = A$$

- **time period, T (s)**
  The time taken for one complete oscillation (imaginary revolution).

- **frequency, f (Hz)**
  The number of oscillations (imaginary revolutions) each second. It is the reciprocal of the time period.

  $$f = \frac{1}{T}$$

- **angular frequency, ω (s⁻¹)**
  This is effectively angular velocity but as the imaginary particle is not real there is no real rotational motion so it is referred to as angular frequency instead. Notice the unit does not contain 'rad'.

  $$\omega = 2\pi f$$

- **maximum velocity, v_{max} (m s⁻¹)**
  This occurs when the object moving in SHM passes through the equilibrium position. It is the same as the tangential velocity of the imaginary particle and is derived using the formula $v = r\omega$ where $r = A$.

  $$v_{max} = A\omega$$

- **maximum acceleration, a_{max} (m s⁻²)**
  This occurs when the object moving in SHM is at maximum displacement. It is the same as the centripetal acceleration of the imaginary particle and is derived using the formulae $a_c = \frac{v^2}{r}$ and $v = r\omega$ where $r = A$.

  $$a_{max} = -A\omega^2$$

By convention, reference circles are drawn with the angle *0 rad* at the '3 o'clock' position and the imaginary particle turns anticlockwise.

Simple trigonometry can be used to determine the displacement, velocity and acceleration of the object moving in SHM for any or time.

Graphing the motion against time reveals the sine and cosine relationships of SHM.

Consider an oscillating mass on a spring at some time, *t*, in its oscillation as shown on the following page.

 ISBN: 9780170368179

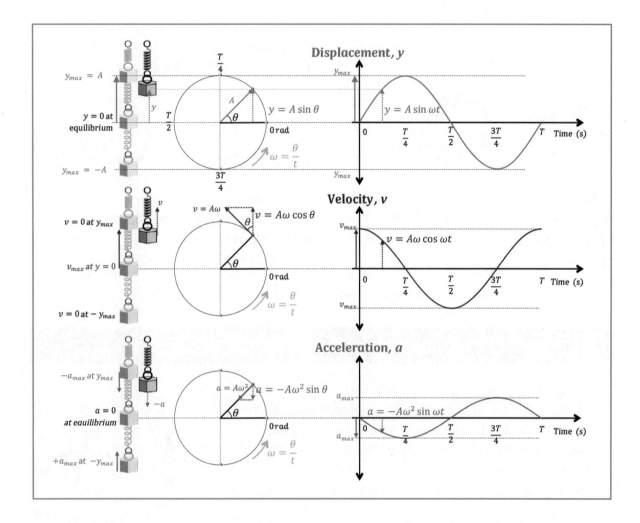

Phasor diagrams

Each of the reference circle diagrams above contains an arrow which shows the maximum size and direction of each quantity. These are called phasors and can be combined on a single diagram to show the relative size and phase relationship between each of the different properties of motion at some point in time, t.

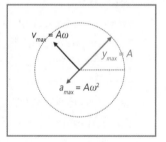

- The velocity phasor leads the displacement phasor by $\pi/2$ (90°).
- The acceleration phasor leads the velocity phasor by $\pi/2$ (90°).
- The acceleration phasor leads the displacement phasor by π (180°).

Solving SHM problems using phasor diagrams

SHM problems can be solved by drawing a reference circle and using simple trigonometry to find the various values, but the general formulae derived from the diagrams above can also be used provided the initial conditions are taken into account.

*Mathematical aside: There can be two answers to an **inverse sine** but your calculator will only ever tell you the answer between $-\dfrac{\pi}{2}$ and $+\dfrac{\pi}{2}$, so check the phasor diagram to make sure that your angle makes sense.*

If the oscillation is measured from either side of the **equilibrium** position ($y = 0$ when $t = 0$ s) then use the **sin**-cos-sin formulae set.

$$y = A \sin \omega t$$
$$v = A\omega \cos \omega t$$
$$a = -A\omega^2 \sin \omega t$$

Think sign:
What is the initial position and motion?
$+y\uparrow, +v\uparrow, -a\downarrow$
$-y\downarrow, -v\downarrow, +a\uparrow$

If the oscillation is measured from the max **displacement** ($y = A/-A$ when $t = 0$ s) then use the **cos**-sin-cos formulae set.

$$y = A \cos \omega t$$
$$v = -A\omega \sin \omega t$$
$$a = -A\omega^2 \cos \omega t$$

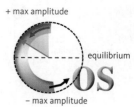

Think sign:
What is the initial position and motion?
$+y\uparrow, -v\downarrow, -a\downarrow$
$-y\downarrow, +v\uparrow, +a\uparrow$

Both sets of formulae reveal that the acceleration $a = -\omega^2 y$

Worked example: Andrew and Bob at the swings

Andrew is playing on a swing. Each swing has an amplitude of 1.5 m and it takes 4.26 s to complete each oscillation. Andrew's friend Bob is standing 1.00 m away from the swing and watches Andrew as he swings from the back to the front as shown in the diagram.
Draw a phasor diagram and hence determine the time taken to reach Bob.

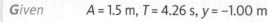

Solution

Where the motion is horizontal, drawing the equilibrium position in the '3 o'clock' position can make it hard to relate the phasor diagram to the picture. It is easier to turn the page ↻ on its side and draw the equilibrium position in the new '3 o'clock' (old '12 o'clock') position then solve as before. The pendulum starts at an amplitude so we need to use the cos-sin-cos set. It is initially in the top half of the phasor diagram so y is +, but Bob is in the lower half of the phasor diagram so $y = -1.00$ m. All SHM equations rely on radians so your calculator must be in radians mode to solve this.

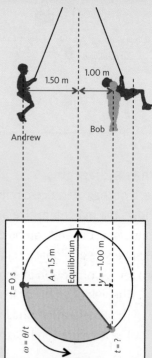

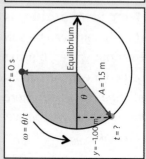

Given	$A = 1.5$ m, $T = 4.26$ s, $y = -1.00$ m
Unknown	$t = ?$
Equations	$f = \dfrac{1}{T}$, $\omega = 2\pi f$, $\omega = \theta/t$, $y = A \cos \omega t$
Substitute	The angular frequency $\omega = \dfrac{2\pi}{T} = \dfrac{2\pi}{4.26} = 1.475$ s⁻¹
	Substituting into cos equation $-1.00 = 1.5 \cos(1.475t)$
	Rearranging gives:
Solve	$1.475t = \cos^{-1}\left(-\dfrac{1.00}{1.5}\right)$ hence $t = \dfrac{2.301}{1.475} = 1.56$ s

 ISBN: 9780170368179

Alternative solution (using basic trig)

Given $\quad$ $A = 1.5$ m, $T = 4.26$ s, $y = -1.00$ m

Unknown $\quad$ $t = ?$

Equations $\quad$ $f = \dfrac{1}{T}$, $\omega = 2\pi f$, $\omega = \theta/t$, $\sin\theta = \dfrac{\text{opp}}{\text{hyp}}$

Substitute $\quad$ The angular frequency $\omega = \dfrac{2\pi}{T} = \dfrac{2\pi}{4.26} = 1.475$ s^{-1}

The displacement phasor travels through a quarter of a circle then a segment. The quarter circle will take a quarter of a time period so
$t_1 = \dfrac{4.26}{4} = 1.065$ s
The angle of the segment $\theta = \sin^{-1}\left(-\dfrac{1.00}{1.5}\right) = 0.7297$ rad
This will take $t = \dfrac{\theta}{\omega} = \dfrac{0.7297}{1.475} = 0.49$ s

Solve $\quad$ Total time taken $t = 1.065 + 0.49 = 1.56$ s

Exercise 4L

1 A mass is oscillating on a spring with an angular frequency of 1.8 s^{-1}. The oscillation has an amplitude of 0.11 m.

a Calculate the maximum velocity of the mass and state where this occurs. (A)

_____

_____

b Calculate the maximum acceleration of the mass and state where this occurs. (A)

_____

_____

c Draw a displacement phasor diagram for the mass in the position shown 0.39 s after it moves up passed the equilibrium position and label it A.

d Find the displacement, velocity and acceleration of the mass at $t = 0.39$ s.

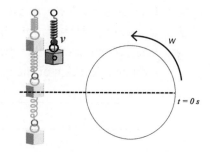

_____

_____

_____

_____

_____

e Draw in the velocity phasor on your diagram and label it v_{max}. (A)

f Draw in the acceleration phasor on your diagram and label it a_{max}. (A)

g Briefly compare the direction of each of your phasors, the motion of the mass and the mathematical sign of the associated answer in part **d**.

_____

_____

_____

_____

_____

2 A buoy in the sea rises and falls with a time period of 5.71 s as the sea waves go past. The buoy travels a total distance of 1.36 m as it rises and falls.

a Calculate the maximum velocity of the buoy. (M)

_____

_____

_____

b Calculate the maximum acceleration. (A)

_____

c Draw a displacement phasor diagram for the buoy in the position shown 1.67 s after it reaches the highest point and label it A. (A)

d Find the displacement, velocity and acceleration of the buoy at $t = 1.67$ s. (A)

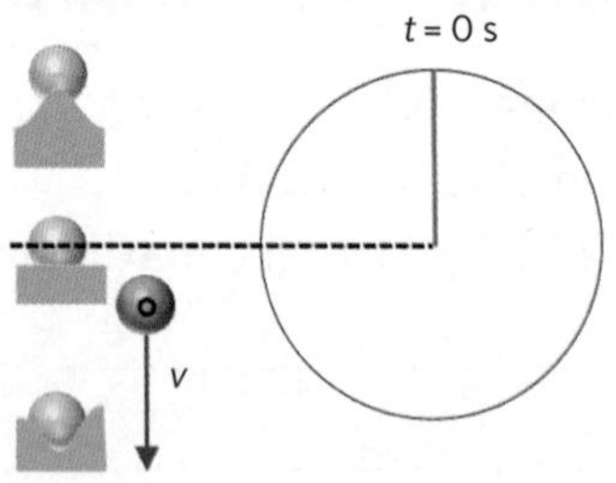

_____

_____

_____

_____

_____

_____

_____

e Draw in the velocity phasor on your diagram and label it v_{max}. (A)

f Draw in the acceleration phasor on your diagram and label it a_{max}. (A)

g Briefly compare the direction of each of your phasors, the motion of the buoy and the mathematical sign of the associated answer in part **d**.

_____

_____

_____

_____

_____

3 A pendulum swings back and forth with a frequency of 0.55 Hz. Starting at the equilibrium position, E, the pendulum bob moves left to the maximum displacement, F, a distance of 14.3 cm then swings back across to the opposite amplitude, G, before being caught at point H, a distance of 8.0 cm from E.

a Show that the angular frequency is 3.46 s⁻¹ and hence calculate

   i the maximum velocity of the pendulum. (M)

       _____

       _____

       _____

   ii the maximum acceleration. (A)

       _____

       _____

       _____

b Draw a displacement phasor diagram for the buoy in the position shown 8.0 cm away from the equilibrium position and label it A. (A)

c Show that it takes 1.6 s to travel E-F-G-H. (M)

Hint: beware the inverse sine – there are two solutions!

_____

_____

_____

_____

d Find the velocity and acceleration of the pendulum at H. (A)

_____

_____

e Draw in the velocity phasor on your diagram and label it v_{max}. (A)

f Draw in the acceleration phasor on your diagram and label it a_{max}. (A)

g Briefly compare the direction of each of your phasors, the motion of the pendulum and the mathematical sign of the associated answer in part **d**.

_____

_____

_____

_____

_____

_____

4 Rebecca and Brittany are sitting in the middle of a pirate ship fairground ride suspended from a pivot 30.0 m above them. The friends are using their mobile phones to measure the speed and velocity of the pirate ship during its motion. The time period of the pirate ship is 11.0 s and it has an amplitude of 15.0 m.

K

15.0 m

$t = 0\,s$

a Show that the angular frequency of the pirate ship is 0.57 s^{-1}. (A)

_____

_____

_____

The friends start taking measurements when the pirate ship is at the left amplitude, K.

• At the moment the pirate ship reaches the outer edge of the tree Rebecca records an acceleration of 4.24 m s^{-2}.

• At the moment it reaches the inner edge of the tree Brittany records a speed of 7.85 m s^{-1}.

b Draw a displacement phasor diagram for the pirate ship showing the point at which the friends reach the outer and inner edges of the tree. (A)

c Calculate the distance of the outer edge of the tree from the equilibrium position. (E)

_____

Hence determine the width of the tree.

_____

_____

_____

_____

Rebecca records an acceleration at the equilibrium position but Brittany says that can't be correct as SHM is proportional to the displacement.

d Discuss Rebecca's data and determine the size of the acceleration. (E)

_____

_____

_____

_____

_____

_____

e What assumption have you made to determine all your answers? Determine whether your assumption is justified. (E)

_____

_____

_____

_____

_____

_____

5

Modern physics

Achievement Standard 91525 (P3.5) requires students to demonstrate understanding by connecting concepts or principles that relate to modern physics. The standard is worth 3 credits and is assessed internally (see the note below).

The Bohr model of the hydrogen atom

- The Bohr model of the hydrogen atom.
- The quantisation of energy.
- Electron transition between energy levels.
- Atomic line spectra.

- The photon.
- Discrete atomic energy levels.
- Ionisation.
- The electron volt.

The photoelectric effect

- The photoelectric effect.

- Wave-particle duality.

Nuclear reactions

- Qualitative description of the effects of the strong interaction and Coulombic repulsion.
- Binding energy and mass deficit.
- Conservation of mass-energy for nuclear reactions.

Relationships (some of these formulae will be part of your course; see the note above)

$$v = f\lambda \qquad E = hf \qquad hf = \phi + E_K \qquad E_K = \frac{1}{2}mv^2 \qquad \Delta E_P = qV$$

$$m_{nucleus} = m_{atom} - m_{electrons} \qquad \Delta m = m_{nucleons} - m_{nucleus} \qquad E = mc^2$$

$$E_n = -\frac{hcR}{n^2} \qquad \frac{1}{\lambda} = R\left(\frac{1}{s^2} - \frac{1}{L^2}\right) \qquad \Delta E = E_{final} - E_{initial}$$

5.0 Quantum theory

Black body radiation

The vibrating molecules and atoms in a warm object will radiate energy, mostly in the form of infrared (IR) radiation. If the object is hot enough it will glow red hot, then white hot and eventually blue hot. A black body is any object that is a perfect emitter (and absorber) of radiation and can be modelled using a blackened ceramic cylinder with a small hole at one end.

The graph below shows the energy radiated at temperatures similar to an oven (400 °C), a filament lamp (2700 °C) and the Sun (5200 °C). As can be seen from a spectrograph, the Sun's light behaves like a black body at the same temperature.

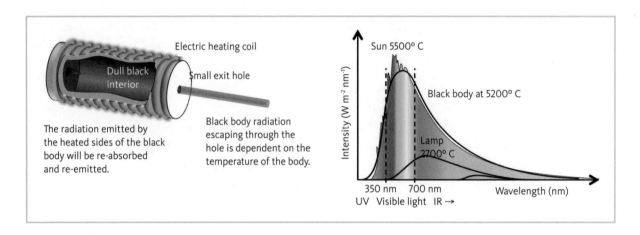

Ultraviolet catastrophe

Towards the end of the 1800s, many attempts were made to explain the observation using **classical theories** based upon mechanics and electromagnetism, but while the formulae could describe the long-wavelength (IR) results, they went catastrophically wrong at short wavelengths (ultraviolet, UV).

Classical physics assumed that the thermal energy of a hot body causes the 'oscillators' (later identified to be atoms and molecules) to vibrate with any amount of energy, resulting in the emission of all wavelengths of electromagnetic waves. As a consequence, classical theory predicted that at short wavelengths an object would emit infinite amounts of energy. This violated the Law of Conservation of Energy and became known as the ultraviolet catastrophe.

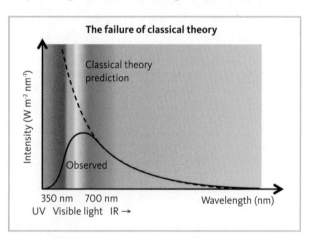

The birth of quantum theory

In 1900, Max Planck showed that an accurate model could be derived on the assumption that:

> The 'oscillators' that emit radiation can only receive or emit discrete quantities or 'packets' of energies known as **quanta**.

Planck stated that each quantum of energy was proportional to the frequency of the oscillation and derived the formula:

$$E = hf$$

which is expressed as:

energy of a quantum (J) = Planck's constant (J s) x frequency of radiation (Hz)

where $h = 6.63 \times 10^{-34}$ J s

Using this quantum theory, short-wavelength, high-frequency oscillators require extremely large amounts of energy to get into the first vibration state. As thermal energy is randomly distributed between the oscillators, the possibility of a high-frequency oscillator receiving enough energy to start vibrating is very low and so the intensity of the radiation emitted at high frequencies drops rapidly to zero, exactly modelling the experimentally observed data as seen in the previous graph.

The electron volt (eV)

The energy associated with a quantum is extremely small and so the electron volt (eV) was developed as the unit of measurement. The electron volt is equal to the kinetic energy gained, ΔE, by an electron with a charge of $q = -1.6 \times 10^{-19}$ C, when it is accelerated through a potential difference of $V = 1$ V. As $\Delta E = qV$ so:

$$\Delta E = 1.6 \times 10^{-19} \times 1 = 1.6 \times 10^{-19} \text{ J}$$

So

$$1 \text{ eV} = 1.6 \times 10^{-19} \text{ J}$$

Exercise 5A

1   Describe a black body. (A)

2   Describe in detail the ultraviolet catastrophe. (M)

3   By considering the graph of the solar spectrum on page **XX**, explain why sunlight appears to be white. (M)

4 Red light has a frequency of 4.50×10^{14} Hz.

    a Calculate the energy of the red light.

    b Calculate the wavelength of the red light.

    c Derive a formula to express energy in terms of wavelength.

5 An oscillator emits a quanta of energy of 5.24×10^{-19} J in the form of violet light. Calculate the frequency of the violet light emitted. (A)

6 Green light has a wavelength of 525 nm. Calculate the quanta of energy of green light. (M)

7 If 1 eV = 1.6×10^{-19} J, how many electron volts are there in 1 joule? Think: do you expect this to be a big or a small number? (A)

8 Calculate the energy of a quanta of ultraviolet light of wavelength 120.0 nm and express your answer in both joules and eV. (M)

9 An oscillator emits a packet of energy of 2.2 eV. Calculate the wavelength of the light emitted and determine the colour of the light. (E)

10 Based on all the previous calculations, which colour(s) of light transfer the most energy? (A)

   ISBN: 9780170368179

Scholarship questions

11 Energy-saving light bulbs are often rated by their 'colour temperature' based on the equivalent temperature of a black body that would produce a similar colour of light. Energy-saving light bulbs are available in two types:

- 'warm white', a golden-yellow-coloured lamp, and
- 'cool daylight', a bluish white light.

By considering the 'colour temperature' of an equivalent black body, discuss whether these descriptions are accurate.

12 A 'cool daylight' energy-saving light bulb produces approximately 3.4×10^{19} photons of visible light each second. If the light bulb has an efficiency of 83%, use reasonable estimates to calculate the power rating of the light bulb.

5.1 The photoelectric effect

In 1887, while testing James Maxwell's theories on electromagnetic waves, Heinrich Hertz observed the **photoelectric effect** for the first time when he noticed that the length of the spark that jumped between two electrodes increased when they were illuminated with ultraviolet (UV) radiation.

Wilhelm Hallwachs used Hertz's observations to conduct an experiment using charged electroscopes. He observed that negative charge could be removed from a negatively charged metal surface by shining UV light on the surface.

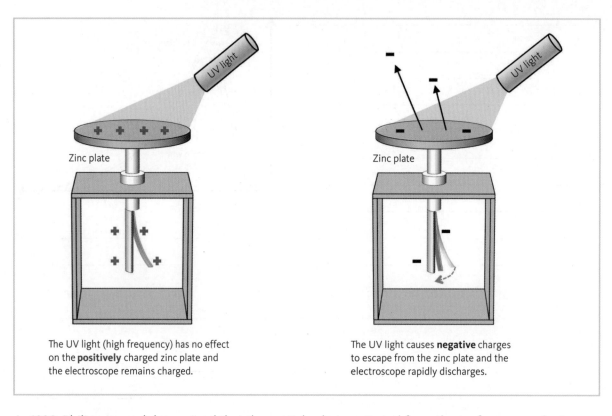

The UV light (high frequency) has no effect on the **positively** charged zinc plate and the electroscope remains charged.

The UV light causes **negative** charges to escape from the zinc plate and the electroscope rapidly discharges.

In 1899, Philipp Lenard determined that the particles being ejected from the surface were electrons. Electrons emitted in this way are referred to as photoelectrons.

Photoelectric effect observations

Further experiments by Philipp Lenard and Robert Millikan led to the following observations:

Observation 1: Photoelectrons are ejected the instant light shines on the metal surface.

Observation 2: The maximum E_K of the photoelectrons is **independent** of the intensity.

 ISBN: 9780170368179

Observation 3: The number of photoelectrons emitted each second is proportional to the intensity (brightness).

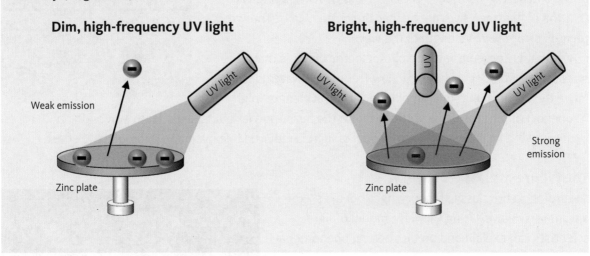

Observation 4: Emission of photoelectrons only occurs if the frequency of the incident light is greater than some **minimum value** called the **threshold frequency**, f_0, which depends upon the type of metal used.

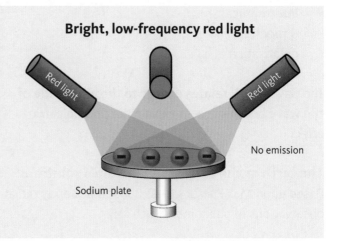

Observation 5: The maximum E_K of the photoelectrons is dependent on the frequency of the incident light and the metal surface being irradiated.

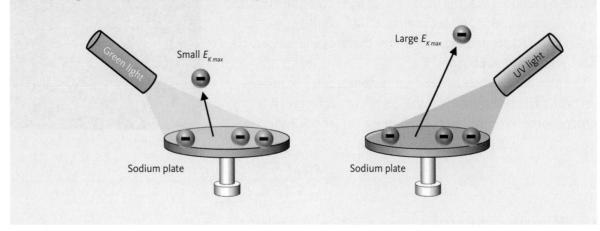

Sea waves analogy

Consider a sea wave approaching the shore where an empty bottle sits partially buried in the sand. The wave model tells us that the higher the wave, the further the bottle will be thrown, regardless of the wavelength of the wave.

But what we observe with the photoelectric effect is that a huge-amplitude, long-wavelength (low-frequency) wave doesn't move the bottle, but a small-amplitude short-wavelength (high-frequency) ripple causes the bottle to fly off the beach and far out to sea. Clearly waves do not behave in this way but light does, so a new model of light is required to explain the photoelectric effect.

The failure of classical theory

According to the classical theory, light behaves like a transverse wave. Using the wave model of light, scientists can explain and predict optical phenomena such as:

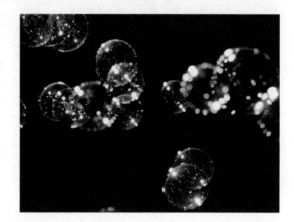

- interference
- diffraction, and
- polarisation.

The wave model **relates energy to the amplitude of the wave** but this fails to explain the photoelectric effect.

The birth of the photon — Einstein's solution

Based upon Max Planck's ideas about black body radiation and light emission, and his own observations of phenomena such as:

- black body radiation
- fluorescence, and
- the photoelectric effect,

Einstein proposed that the behaviour of light could be explained if it was considered to be concentrated into a finite number of individual packets or quanta of energy that move like a particle, later referred to as **photons**.

The photons can only be produced or absorbed as complete units and Einstein stated that the energy of each photon is proportional to the frequency according to Planck's equation:

$$E = hf$$

Einstein proposed that when a photon collides with an electron, it must:
- either be reflected with no loss of energy
- or be completely absorbed giving up all its energy to the electron.

Classical wave theory versus particle theory

The table below relates the classical theory predictions to the observed outcomes of the photoelectric effect.

Classical wave theory predictions		Photoelectric observation		Particle theory predictions
All frequencies of light should cause emission provided the amplitude is large enough to transfer sufficient energy to the electrons on the metal surface.	✗	**Observation 4:** No emission occurs below the threshold frequency, f_0 regardless of the intensity of the light.	✓	Lower frequencies of light have less energy so will only be able to eject electrons if the photon's energy is greater than the work that must be done to overcome the forces binding the electron to the surface.
As light from a lamp travels away, the energy becomes increasingly spread out and the **amplitude** of the wave **decreases.** As each electron should receive an equal share of the energy, a significant amount of time will need to pass before any single electron gains sufficient energy to escape the forces binding it to the metal surface. Low-intensity light will have a smaller amplitude, and therefore deliver only a small amount of energy spread over the wavefront.	✗	**Observation 1:** Photoelectrons are ejected the instant the surface is irradiated even by low-intensity light. *(Provided the frequency of the incident light is above the threshold frequency.)*	✓	As light from a lamp travels away, the energy becomes increasingly spread out but the energy of each individual photon remains the same. Provided that $f > f_0$, even a single photon (very low intensity) can deliver sufficient energy to an individual electron causing it to be ejected from the surface almost instantly.
Brighter light with a greater intensity will have a greater amplitude and carry more energy, so should cause the maximum E_K of the photoelectrons to increase.	✗	**Observation 2:** The maximum E_K of the photoelectrons is **independent** of the intensity.	✓	For a particular frequency, brighter, more intense light still has the same energy, so the maximum E_K will remain the same.
Brighter light should also increase the number of photoelectrons that can overcome the binding forces holding them in place.	✓	**Observation 3:** Greater intensity results in more photoelectrons being ejected.	✓	Increasing the intensity should increase the number of photons striking the surface and therefore increase the number of photoelectrons ejected.
The maximum E_K should be independent of the frequency of the incident light.	✗	**Observation 5:** The maximum E_K of the photoelectrons is dependent on the frequency of the incident light.	✓	Increasing the frequency of the incident light should increase the energy of each photon and hence increases the maximum E_K of the ejected photoelectrons.

Work function, φ

To explain the threshold frequency, f_0, Einstein proposed that some of the energy delivered by the photon was required to release the electron from the surface by doing work against the binding forces that hold it. The **work function**, φ, describes the amount of energy required to free the electron from the surface. For a particular material:

If the frequency of the light $f < f_0$	If the frequency of the light $f = f_0$	If the frequency of the light $f > f_0$
then the energy delivered by the photon is **less than** the work function of the surface.	then the energy delivered by the photon is equal to the work function.	then the energy delivered by the photon is **greater than** the work function.
$$hf < \varphi$$	$$hf_0 = \varphi$$	$$hf > \varphi$$
Photon energy $E < hf_0$ 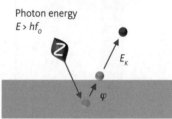	Photon energy $E = hf_0$	Photon energy $E > hf_0$
Insufficient energy to overcome the binding forces holding the electron in place.	The electron has absorbed sufficient energy to overcome the work function of the surface but there is no energy remaining to provide the electron with E_K to escape the surface and so it eventually falls back into the surface.	The electron has absorbed sufficient energy to overcome the work function of the surface and the remaining energy is available as E_K to escape the surface. (Note: Most electrons lose some E_K before emission due to collisions with other atoms on the way to the surface.)

Einstein determined that the **maximum kinetic energy** available to any photoelectron is given by the equation:

$$E_{K\,max} = hf - \varphi$$

which states that:

maximum kinetic energy (J) = energy supplied by the incident photon (J) – work function (J)

Einstein's photoelectric equation

  ISBN: 9780170368179

Worked example: Sodium and the photoelectric effect

Freshly cut sodium has a threshold frequency of 5.55×10^{14} Hz and photoelectrons can be ejected from the surface of the metal by irradiating it with light of wavelength 520 nm. By first determining the work function, calculate the maximum velocity of the photoelectrons. Use $m_e = 9.11 \times 10^{-31}$ kg, $h = 6.63 \times 10^{-34}$ J s, $c = 2.998 \times 10^8$ m s^{-1}. (E)

Solution

The work function, φ, can be determined by applying the knowledge that $\varphi = hf_0$.

Given	$h = 6.63 \times 10^{-34}$ J s, $f_0 = 5.55 \times 10^{14}$ Hz
Unknown	$\varphi = ?$
Equations	$E = hf$, which becomes $\varphi = hf_0$ at the threshold frequency.
Substitute	$\varphi = 6.63 \times 10^{-34} \times 5.55 \times 10^{14}$
Solve	$\varphi = 3.68 \times 10^{-19}$ J

The maximum kinetic energy can be determined by applying Einstein's photoelectric equation, but first it is important to convert the wavelength from nm to m to find the frequency of the light incident on the surface.

Given	$h = 6.63 \times 10^{-34}$ J s, $c = 2.998 \times 10^8$ m s^{-1}, $m_e = 9.11 \times 10^{-31}$ kg, $\lambda = 520$ nm $= 520 \times 10^{-9}$ m
Unknown	$v_{max} = ?$
Equations	Combining $E_K = \frac{1}{2} mv^2$ and $E_K = hf - \varphi$, $f = c/\lambda$ gives: $$\frac{1}{2} mv^2 = \frac{hc}{\lambda} - \varphi$$
Substitute	$$\frac{1}{2} \times 9.11 \times 10^{-31} \times v^2 = \frac{6.63 \times 10^{-34} \times 2.998 \times 10^8}{520 \times 10^{-9}} - 3.68 \times 10^{-19}$$
Solve	$4.555 \times 10^{-31} \times v^2 = 3.822 \times 10^{-19} - 3.68 \times 10^{-19}$ $$v^2 = \frac{1.42 \times 10^{-20}}{4.555 \times 10^{-31}} = 3.13 \times 10^{10}$$ $v_{max} = \sqrt{3.13 \times 10^{10}} = 176\,843$ m s^{-1} = 180 km s^{-1} (2 sf)

Exercise 5B

Use: $h = 6.63 \times 10^{-34}$ J s; $c = 2.998 \times 10^8$ m s^{-1}

1 Describe the photoelectric effect. (A)

_____

_____

2 Describe the work function of a material and relate it to the threshold frequency. (M)

_____

_____

3 The threshold frequency of caesium is 5.16×10^{14} Hz.

    a Calculate the work function for caesium. (A)

    _____

    _____

    b Give the answer in electron volts. (A)

    _____

    _____

4 The work function of aluminium is 4.12 eV. Calculate the threshold frequency. (M)

_____

_____

_____

5 When ultraviolet light of frequency 8.00×10^{14} Hz is incident on a surface, the photoelectrons are ejected with a range of speeds. Explain why photoelectrons do not all leave the surface at the same speed. (M)

_____

_____

_____

_____

6 Explain the effect of **increasing the intensity** of incident light on the emission of photo-electrons from a metal surface. In your answer you should consider the number and speed of the emitted photoelectrons. (M)

_____

_____

_____

_____

_____

_____

_____

 ISBN: 9780170368179

7   Explain the effect of **decreasing wavelength** of incident light on the emission of photo-
    electrons from a metal surface. In your answer you should consider the number and speed of
    the emitted photoelectrons. (E)

    _____

    _____

    _____

    _____

    _____

    _____

    _____

    _____

    _____

    _____

    _____

8   Gold with a work function of 5.10 eV is irradiated with UV light of wavelength 166.67 nm.

    a   Show that the maximum kinetic energy of photoelectrons emitted from gold is 2.4 eV. (M)

        _____

        _____

        _____

        _____

        _____

        _____

    b   Given that an electron has a mass of m_e = 9.11 x 10^{-31} kg, calculate the maximum speed of
        photoelectrons emitted from the gold. (A)

        _____

        _____

        _____

        _____

        _____

        _____

9 Calculate the wavelength of light necessary to cause photoelectrons to be emitted from a lithium sheet with a speed of 750 km s⁻¹. The work function of lithium is 2.90 eV and the mass of an electron is $m_e = 9.11 \times 10^{-31}$ kg. (E)

_____

_____

_____

_____

_____

_____

_____

_____

_____

Scholarship questions

10 Discuss the idea of wave-particle duality by comparing and contrasting the wave and particle models of light with experimental evidence.

_____

_____

_____

_____

_____

_____

_____

_____

_____

_____

_____

_____

_____

_____

_____

11 Discuss the differences between a beta particle emission and photoelectron emission.

_____

_____

_____

_____

_____

_____

_____

_____

_____

_____

_____

_____

_____

Millikan's experiment

Millikan carried out a series of experiments to test Einstein's theory by shining light of a known frequency on a test metal (referred to as the emitter) and collected the photoelectrons emitted from the surface.

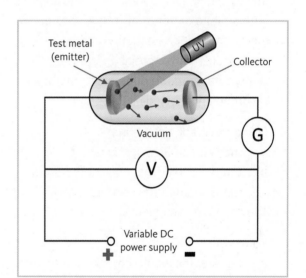

The photoelectrons that reached the collector plate created a tiny current that was measured by a galvanometer.

By applying a small positive potential to the emitter, the emitted photoelectrons experienced an electrostatic force of attraction pulling them back towards the emitter.

The work done by this force caused photo-electrons with a low E_K to be pulled back, and only those photoelectrons with a higher E_K made it to the collector plate. As a consequence, the current decreased.

Millikan gradually increased the positive potential on the emitter until the current fell to zero, allowing him to determine the voltage required to just stop the photoelectrons with the maximum kinetic energy. At this **stopping potential** V, the **work done**, ΔE, on the photoelectrons by the electrostatic force is equal to the maximum kinetic energy of the photoelectrons.

$$\Delta E_{stopping} = E_{K\,max}$$

Combining this with Einstein's photoelectric equation gives:

$$\Delta E_{stopping} = hf - \varphi$$

A graph of kinetic energy against frequency reveals a linear relationship, which is of the form '$y = mx + c$'. Different metals have different work functions and threshold frequencies, so the y and x intercepts change. However, the gradient remains unchanged, as this is Planck's constant, h.

For an electron, the stopping potential, V_s, is given by:

$$V_s = \frac{\Delta E_{stopping}}{e}$$

The photoelectric equation can also be expressed as:

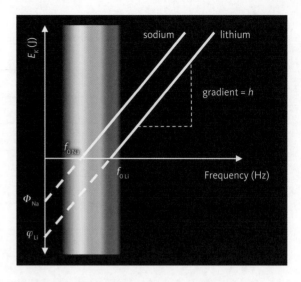

$$V_s = \frac{h}{e}f - \frac{\varphi}{e}$$

which enables the threshold frequency and the work function to be determined by changing the frequency and measuring the stopping potential. See the following worked example.

Worked example: Photocathode

A sheet of freshly cut rubidium is sealed in an evacuated glass container and irradiated with a range of different frequencies of electromagnetic radiation. At different frequencies the potential difference required to stop the photoelectrons was recorded and plotted on the graph.

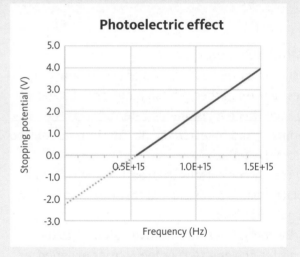

Using the graph determine:

a   the value of Planck's constant

b   the threshold frequency of rubidium

c   the work function of rubidium.

The charge on an electron $e = 1.6 \times 10^{-19}$ C

Solution

a   Beware: The graph shows **stopping potential** vs frequency, so to understand the gradient of the graph, combine the photoelectric formula with the charge-energy formula.

Given	Data points (1.5×10^{15}, 4.00) and (5.5×10^{14}, 0.00), $e = 1.6 \times 10^{-19}$ C
Unknown	$h = ?$
Equations	$\Delta E_{stopping} = hf - \varphi$ and $\Delta E_{stopping} = eV_s$
Substitute	$eV_s = hf - \varphi$ so $V_s = \frac{h}{e}f - \frac{\varphi}{e}$ so the gradient of the graph $m = \frac{h}{e}$
	The gradient of the graph $m = \dfrac{4.00 - 0.00}{1.5 \times 10^{15} - 5.5 \times 10^{14}} = 4.2 \times 10^{-15}$ V s
Solve	Hence $\dfrac{h}{e} = 4.2 \times 10^{-15}$
	So $h = 4.2 \times 10^{-15} \times 1.6 \times 10^{-19} = 6.7 \times 10^{-34}$ J s

b The threshold frequency can be determined from the *x*-axis intercept.

Given	Data point $(5.5 \times 10^{14}, 0.00)$
Unknown	$f_0 = ?$
Equations	$V = \dfrac{h}{e}f - \dfrac{\varphi}{e}$
Substitute	When $V = 0$ V then $h\,f = \varphi$ and as $\varphi = hf_0$ so $f = f_0$
Solve	Hence $f_0 = 5.5 \times 10^{14}$ Hz

c The work function can be determined by extrapolating down to the *y*-axis intercept (or by using the relationship $\varphi = hf_0$, but this relies on the previous calculation being correct!).

Given	Data point $(0.00, -2.25)$
Unknown	$\varphi = ?$
Equations	$V = \dfrac{h}{e}f - \dfrac{\varphi}{e}$
Substitute	When $f = 0$ Hz then $V = -\dfrac{\varphi}{e}$ and so $\varphi = -eV$
Solve	Hence $\varphi = -1.6 \times 10^{-19} \times -2.25 = 3.6 \times 10^{-19}$ J

Saturation current and the photoelectric effect

When an increasingly **negative** potential is applied to the **collector**, it decreases the number of photoelectrons that can reach the collector. The stopping potential, V_s, is the voltage at which the current falls to zero.

Applying a **zero** potential to the collector allows a photoelectric current to flow, however not all the photoelectrons are emitted perpendicular to the surface and so don't travel towards the collector, hitting the glass sides instead.

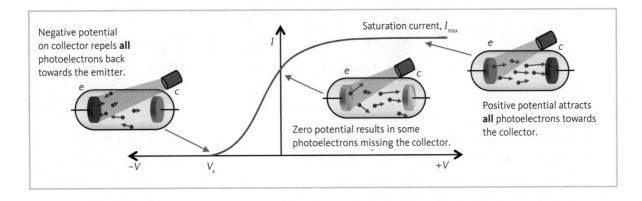

Negative potential on collector repels **all** photoelectrons back towards the emitter.

Saturation current, I_{max}

Zero potential results in some photoelectrons missing the collector.

Positive potential attracts **all** photoelectrons towards the collector.

Applying an increasingly **positive** potential to the collector attracts more of the emitted photoelectrons towards the collector and causes the current to increase. When all the emitted photoelectrons are striking the collector, a maximum **saturation current**, I_{max}, is reached.

The saturation current is limited by the number of photoelectrons being emitted, which in turn is limited by:

- intensity — greater intensity means more photons striking the emitter each second, so more photoelectrons per second and a greater current.
- frequency — when a photon is absorbed by an electron that is deeper within the surface, the emitted electron may lose all its E_K before it reaches the surface due to collisions with other electrons. Higher frequency photons provide more energy, so electrons deeper within the surface will also escape and contribute to the total current.

The current is dependent on the number of photoelectrons leaving the emitter each second, not how fast they travel to the collector. Therefore the current is independent of the speed.

Wave-particle duality

The ability of wave theory to explain and predict reflection, refraction, diffraction, interference and polarisation appeared to confirm that light was indeed a wave motion. However, observations of the photoelectric effect, fluorescence and the ultraviolet catastrophe clearly demonstrate that light also behaves like a particle.

Light is neither a wave nor a particle, but its behaviour can be modelled using either the wave model or the particle model as appropriate.

Exercise 5C

Use: charge on electron $e = 1.602 \times 10^{-19}$ C, $h = 6.626 \times 10^{-34}$ J s, $c = 2.998 \times 10^8$ m s^{-1}
Questions **1–3** are based on the photocathode circuit shown below.

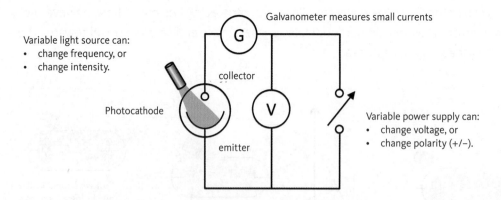

1   Solar cells use the photoelectric effect to generate electricity. Megan decides to test calcium as a possible emitter material for use in a solar cell and she illuminates it with a monochromatic light. The lamp initially emits red light, but as the wavelength is decreased it changes colour until it is finally producing violet light. No current is recorded by the galvanometer until the light turns greeny-blue with a wavelength of 432 nm.

a Calculate the work function of the calcium emitter. (A)

_____

_____

b Explain why no current is recorded in the circuit until the wavelength falls below 432 nm. (M)

_____

_____

_____

_____

While illuminating the calcium with violet light of wavelength 350 nm, the intensity of the light is gradually varied and the current recorded as shown in the graph below.

c Explain why increasing the intensity increases the current. (M)

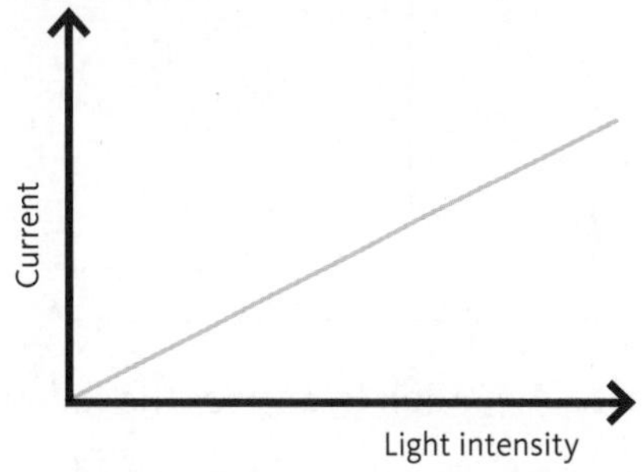

_____

_____

_____

_____

_____

_____

A positive potential is applied to the collector plate and the light intensity test is repeated.

d Draw a line on the graph above to show how the current varies with the light intensity when a positive potential is applied to the collector plate. Explain your answer. (M)

_____

_____

_____

_____

_____

While illuminating the calcium with high-intensity violet light of frequency 8.566×10^{14} Hz, an increasing positive potential is now applied to the emitter until the current falls to zero.

e Name the value of the potential at which this occurs and explain why the current falls to zero. (M)

_____

_____

_____

f Calculate the size of the potential required to reduce the current to zero. (A)

_____

_____

_____

2 The emitter of a photocathode is made of barium and is illuminated with a monochromatic light resulting in the emission of photoelectrons. The voltage across the photocathode is increased until the positive charge on the emitter plate reduces the current to zero, and the frequency and potential difference recorded. The frequency of the light is then increased and the new stopping potential measured as shown in the graph.

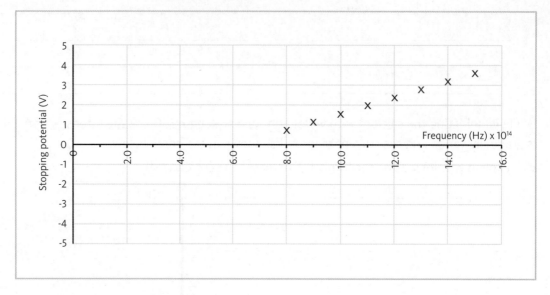

Use the graph to answer the following.

a Determine the threshold frequency. (A)

_____

b Show that the work function of barium is 4.2×10^{-19} J. (M)

_____

_____

_____

c Find the experimental value for Planck's constant; give a unit with your answer. (M)

_____

_____

_____

_____

The barium emitter was then replaced with an aluminium emitter with a stopping potential of 4.10 V and the experiment was repeated.

d On the graph above, draw a line to show the results you would expect from the experiment, explain your reasons and determine the threshold frequency for aluminium. (E)

_____

_____

_____

_____

_____

_____

3 Using a fixed frequency and intensity of the lamp, the voltage was increased from a negative potential to a positive potential across the photocathode. The results for the current in the circuit are shown in the graph below.

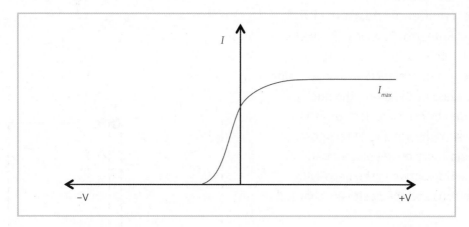

a On the graph above, draw a graph line to show the results you would expect to get if the experiment was repeated with incident radiation of the same frequency but with a brighter light (greater intensity) and label it 'Greater intensity'. Discuss your graph. (E)

_____

_____

_____

_____

_____

_____

_____

_____

_____

b On the graph above, draw a graph line to show the results you would expect to get if the experiment was repeated with incident radiation of the same brightness but with a higher frequency, and label it 'Higher frequency'. Explain your graph. (E)

_____

_____

_____

_____

_____

_____

_____

_____

_____

_____

5.2 The Bohr model of the hydrogen atom

Models of the atom

Rutherford's model of the atom (1911)

In 1911, Ernest Rutherford proposed a nuclear model in which the positive charge is concentrated in a dense nucleus that contains 99.9% of the mass of the atom. The nucleus was surrounded by orbiting electrons that occupied most of the volume of the atom. They were held there by the electrostatic force of attraction between the negatively charged electrons and the positively charged nucleus.

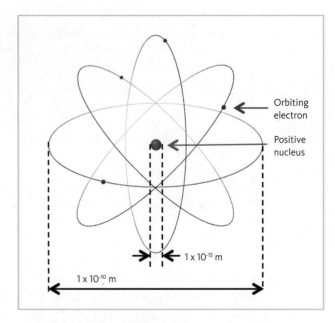

Rutherford's model of the atom was consistent with his observations of the alpha particle scattering experiment, but it was inconsistent with classical physics theories which stated that an accelerating electron must emit electromagnetic radiation.

As Rutherford's orbiting electrons were moving in circular orbits, they were experiencing a centripetal acceleration and as a result should continuously emit radiation, causing them to lose energy, slow down, and spiral into the nucleus, collapsing the atom.

Bohr model of the hydrogen atom (1913)

In 1913, Niels Bohr developed Rutherford's model by using ideas taken from quantum theory and applying two assumptions.

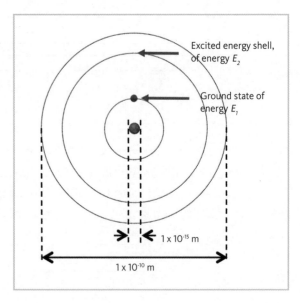

- Similar to Rutherford, Bohr stated that electrons move in circular orbits around the nucleus due to the action of the electrostatic force. The greater the radius of the orbit, the greater the energy (potential plus kinetic) of the electron.

- However, Bohr hypothesised that an electron could only travel in certain 'allowed orbits' in which:

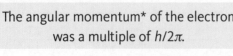

  The angular momentum* of the electron was a multiple of $h/2\pi$.

  * See Angular momentum, page 164.

By limiting the angular momentum to specified amounts rather than a continuous range of values, Bohr had effectively quantised atomic structure. Bohr stated that while an electron was in one of these orbits, it did not radiate electromagnetic radiation and had a defined **energy level** specific to each orbit. (See Extension concepts for scholarship: de Broglie's hypothesis for a more in-depth discussion of energy levels, page 217.)

- Bohr proposed that an electron can 'jump', or **transition**, from a higher energy orbit, E_2, to a lower energy orbit, E_1. In doing so the electron releases a **quantum** of energy equal to the energy difference:

$$E = E_2 - E_1$$

in the form of electromagnetic radiation of frequency, f, as given by Planck's equation,

$$E = hf$$

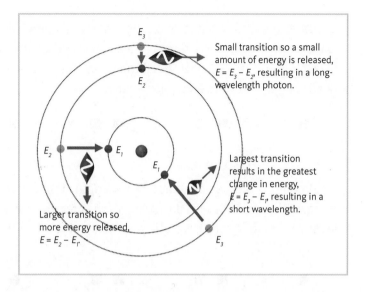

Small transition so a small amount of energy is released, $E = E_3 - E_2$, resulting in a long-wavelength photon.

Larger transition so more energy released, $E = E_2 - E_1$.

Largest transition results in the greatest change in energy, $E = E_3 - E_1$, resulting in a short wavelength.

Bohr's model of the hydrogen atom explained the origin of electromagnetic radiation from within atoms and accurately predicted the frequencies of the light given off by hydrogen when it is heated (see Atomic spectra, page 224).

Extension concepts for scholarship — De Broglie's hypothesis

Bohr's model failed to generate accurate data for the light given off by atoms with more than one electron. Bohr had no theoretical basis for the arbitrary assumption that angular momentum is quantised and his mathematical derivation relied on quantities, such as the radius, that could not be confirmed experimentally, however it introduced the fundamental ideas that are still used in modern **wave mechanics** today.

Based upon the ideas of the wave-particle duality of light, Louis de Broglie (pronounced *de broy*) hypothesised in 1924 that particles of matter, such as electrons, may also exhibit wave-like behaviours. De Broglie reasoned that as the energy of a photon is given by $E = hf$ and Einstein's mass-energy relation states that $E = mc^2$, so $hf = mc^2$.

As $f = c/\lambda$, we have $\frac{hc}{\lambda} = mc^2$, which simplifies to $\frac{h}{\lambda} = mc$, which is the momentum of the photon, and as $p = mc$ we have:

$$\lambda = \frac{h}{p}$$

The de Broglie wavelength

De Broglie proposed that any particle of relativistic mass m moving with momentum p would exhibit wave-like behaviour with a de Broglie wavelength λ. In 1927, his theory was confirmed when Clinton Davisson and Edmund Germer successfully diffracted electrons using a crystal of nickel resulting in a diffraction pattern similar to that produced by light but with a wavelength matching de Broglie's prediction.

Electron waves and the quantised atom

Rutherford introduced the idea of electrons as particles orbiting the nucleus. But in order to explain atomic spectra, Bohr developed this idea by including the requirement that electrons could only exist in certain discrete orbits (see The Bohr model of the atom, page 216). De Broglie's idea of matter waves provides a simple analogy as to why the electron can only exist in these discrete orbits.

Consider the waves that form on the string of a guitar when it is plucked. Only multiples of half wavelengths will fit the string, and these will constructively interfere resulting in a standing wave (see Waves on strings, page 73). If the wavelength does not fit the string, then the waves will destructively interfere and quickly cancel out.

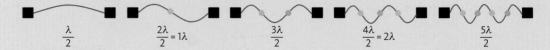

$\frac{\lambda}{2}$ $\qquad$ $\frac{2\lambda}{2} = 1\lambda$ $\qquad$ $\frac{3\lambda}{2}$ $\qquad$ $\frac{4\lambda}{2} = 2\lambda$ $\qquad$ $\frac{5\lambda}{2}$

De Broglie considered only those wavelengths that formed a whole number of wavelengths (that is, have the same phase at each end) so the electron wave could join back up with itself. This leads to a set of discrete wavelengths that will form on the string as shown below.

These waves are now bent to fit over the circular orbits, modelling the electron waves of the ground state and excited states of an atom as shown below.

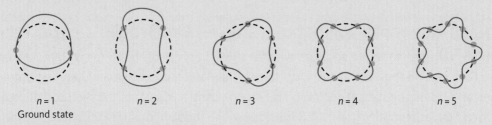

$n = 1$ $n = 2$ $n = 3$ $n = 4$ $n = 5$

Ground state

In the excited states the electrons have:

- a shorter de Broglie wavelength
- which means a higher frequency
- hence more energy, and
- greater momentum, as $p = \frac{h}{\lambda}$.

De Broglie's quantum condition was that only those wavelengths that 'fit' the circumference a whole number of times will produce a stable orbit. If the **quantum number**, **n**, is equal to the number of complete wavelengths that fit the circumference of the orbit, then $2\pi r = n\lambda$. But according to de Broglie, $\lambda = \frac{h}{mv}$ so $2\pi r = \frac{nh}{mv}$, which can be rearranged to give:

$$\text{angular momentum} = mvr = n\frac{h}{2\pi}$$

which reveals that the angular momentum of the electron is also quantised (see Angular momentum, page 164), a condition that was originally proposed by Bohr.

 ISBN: 9780170368179

Discrete atomic energy levels based on Bohr's model

According to wave mechanics, the electrons exist in **energy levels** denoted by their **quantum number**, $n = 1, 2, 3, \ldots \infty$, which have discrete amounts of energy. All atoms of the same element have the same unique set of discrete energy levels, a property that enables the element to be easily identified, even across the far reaches of space. The energy of each level can be calculated using wave mechanics, and the values are typically represented using an energy level diagram. The energy level diagram for hydrogen is shown below with a sample set of transitions from the $n = 4$ level.

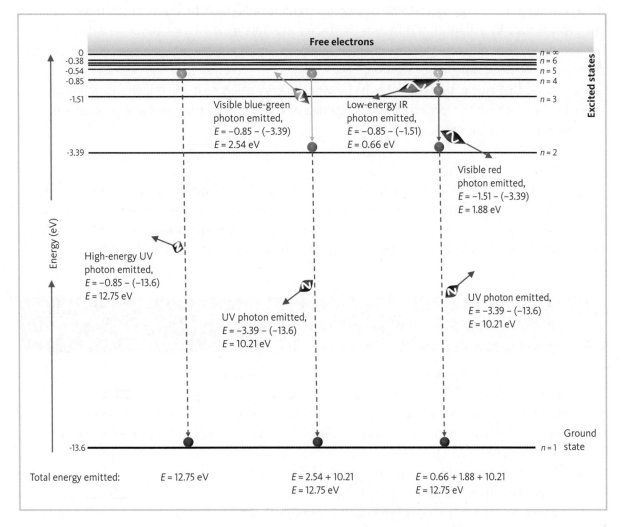

A free electron at rest outside an atom, that is, $n = \infty$, is considered to have zero energy. When it 'falls' into an atom the electron loses energy, which is given out as a photon of energy equal to the difference between the energies of the two levels.

$$E = E_{higher} - E_{lower}$$

The energy levels are usually given in eV (due to their small size) and are quoted as negative, as they represent the amount of energy that must be **supplied** to take an electron from that level and completely remove it to zero energy outside the atom.

The most stable state for the atom is when the electron reaches the lowest energy level called the **ground state**, which is given the **quantum number** $n = 1$. For the hydrogen atom, the ground state has an energy of −13.6 eV.

If the atom gains exactly the right amount of energy by:

- heating
- collisions with other particles
- electricity, or
- absorbing electromagnetic radiation,

the electron will be promoted to a higher energy level and the atom is now in an **excited state**.

Excited states are unstable and after a short time the electron will fall back down to the ground state, emitting the energy as a photon of electromagnetic radiation. The electron can travel down to the ground state by a number of different routes, for example $n = 4 \rightarrow n = 1$, or $n = 4 \rightarrow n = 3 \rightarrow n = 1$. Each route involves different discrete changes in energy and, consequently, the emission of different frequencies of electromagnetic radiation.

Bohr stated that an electron cannot exist between energy levels so it will only transition from one level to another if the 'packet' of energy gained or lost is exactly the same size as the energy difference between the two levels.

Ionisation

To ionise a hydrogen atom, that is, remove an electron completely from the ground state $n = 1 \rightarrow n = \infty$, work must be done to overcome the electrostatic force of attraction between the negative electron and the positive nucleus. In a hydrogen atom, 13.6 eV of energy would need to be supplied to completely remove an electron from the ground state and ionise the atom.

 ## Worked example: Frequency of emission for a transition from $n = 5$ to $n = 2$

Heating monatomic hydrogen gas causes the atoms to become excited. An electron that has been promoted to the fifth energy level eventually transitions back down to the second level as shown in the diagram and emits a photon of violet light. Planck's constant $h = 6.63 \times 10^{-34}$ J s.

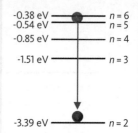

a Show that the energy change as a result of the transition from the fifth to the second level is 4.6×10^{-19} J.

b Determine the frequency of the emitted photon.

c State and explain how much energy would be required to remove an electron in the $n = 2$ level completely from the atom.

Solution

a This is a 'Show that' question, so the value will need to be used later so highlight it now!

Given	$E_5 = -0.54$ eV and $E_2 = -3.39$ eV
Unknown	$E_{5 \rightarrow 2} = ?$
Equations	$E_{5 \rightarrow 2} = E_5 - E_2$
	$E_{joules} = E_{eV} \times 1.6 \times 10^{-19}$
Substitute	$E_{5 \rightarrow 2} = -0.54 - (-3.39) = 2.85$ eV
Solve	$E_{joules} = 2.85 \times 1.6 \times 10^{-19} = 4.56 \times 10^{-19} = 4.6 \times 10^{-19}$ J (2 sf)

b All frequency calculations must be carried out using joules, not eV.

> *Given* $E_{5\to2} = 4.56 \times 10^{-19}$ J and $h = 6.63 \times 10^{-34}$ J s
>
> *Unknown* $f_{5\to2} = ?$
>
> *Equations* $E = hf$
>
> *Substitute* $f_{5\to2} = \dfrac{4.56 \times 10^{-19}}{6.63 \times 10^{-34}} = 6.88 \times 10^{14}$ Hz
>
> *Solve* $f_{5\to2} = 6.88 \times 10^{14} = 6.9 \times 10^{14}$ Hz (2 sf)

c The $n = 2$ energy level has an energy of -3.39 eV. This represents the amount of energy required to remove the electron from this level to infinity, leaving the atom as a positive ion.

Exercise 5D

Constants: $h = 6.626 \times 10^{-34}$ J s, $e = 1.602 \times 10^{-19}$ C, $c = 2.998 \times 10^{8}$ m s^{-1}

1 Describe Rutherford's model of the atom and explain why it had to be modified. (M)

_____

_____

_____

_____

_____

2 Explain how Bohr quantised the atom. (M)

_____

_____

_____

_____

_____

_____

_____

_____

3 Explain why energy levels are negative. (M)

_____

_____

_____

_____

_____

4   Explain what will happen if a photon with an energy that doesn't match a transition is incident on an H atom. (M)

_____

_____

_____

_____

_____

_____

_____

5   A high potential is placed across a low-pressure gas tube containing monatomic hydrogen gas causing the atoms to become excited. An electron that has been promoted to the fourth energy level eventually transitions back down to the second shell, as shown in the diagram, and emits a photon of violet light.

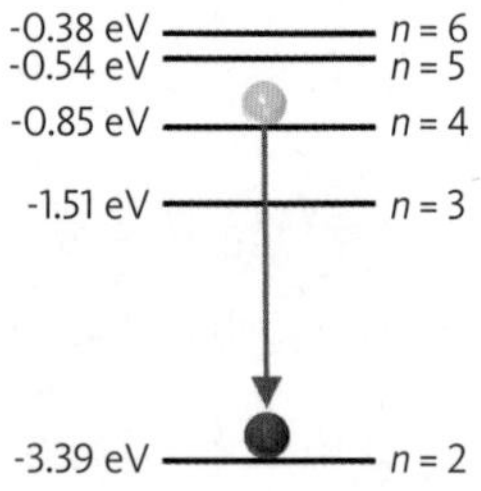

-0.38 eV	n = 6
-0.54 eV	n = 5
-0.85 eV	n = 4
-1.51 eV	n = 3
-3.39 eV	n = 2

a   Show that the energy change as a result of the transition from the fourth to the second level is 4.07×10^{-19} J. (A)

_____

_____

_____

b   Determine the wavelength of the emitted photon. (A)

_____

_____

6   Ultraviolet light of wavelength 94.95 nm is incident on a low-pressure hydrogen gas cloud causing an electron to transition from the ground state of energy -13.6 eV up to a higher energy level. Using the diagram on page 219, determine to which level the electron transitions. (E)

_____

_____

_____

_____

_____

_____

_____

_____

_____

_____

   ISBN: 9780170368179

Scholarship question

7 Optical microscopes are limited in their ability to view really fine detail, as they cannot separate objects closer than 2×10^{-7} m. However, a transmission electron microscope (TEM) can view much finer detail using electron beams instead of light.

A TEM accelerates electrons through a potential difference of 50 kV prior to striking the object being studied. By considering the de Broglie wavelength of an electron of mass $m = 9.11 \times 10^{-31}$ kg that has been accelerated in the microscope, estimate the minimum separation of two objects with the TEM and explain the limiting factor. Charge on an electron, $q = 1.60 \times 10^{-19}$ C.

_____

_____

_____

_____

_____

_____

_____

_____

_____

_____

_____

_____

_____

_____

_____

_____

_____

_____

_____

_____

_____

Spectra

Continuous spectra

The atoms in a solid, liquid or high-pressure gas are so close together that they interact with each other causing the electrons to have a continuous range of energies.

When excited, these electrons will radiate electromagnetic radiation over a continuous range of wavelengths with the intensity of different colours dependent only upon the temperature of the object (see Black body radiation curve of the sun, page 196).

For example, a hot filament lamp produces white light, which will form a complete rainbow when viewed through a prism or diffraction grating (see Diffraction grating, page 65).

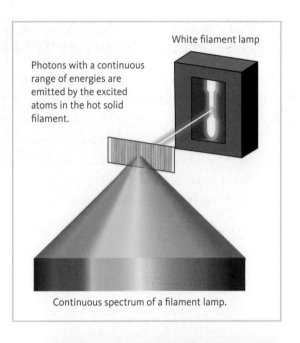

White filament lamp

Photons with a continuous range of energies are emitted by the excited atoms in the hot solid filament.

Continuous spectrum of a filament lamp.

Atomic line spectra

When a monatomic (single-atom) low-pressure gas is excited, it will emit electromagnetic radiation due to the electron transitions taking place within the atoms of the gas.

The frequency of the light emitted depends on the difference in the energy levels through which the electrons transition. As each element has different energy levels, the light emitted by each gas is unique to the element.

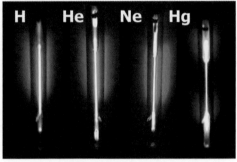

H, He, Ne and Hg gas discharge tubes.

Optical emission spectra

If visible light radiated directly from a source such as:

- a flame
- a low-pressure gas discharge tube, for example a neon lamp, or
- a gaseous nebula in space

is passed through a prism or diffraction grating, then an emission line spectrum is produced.

The energy levels in hydrogen have discrete values, so when the electron in the hydrogen atom transitions down to the $n = 2$ energy level, it releases energy in discrete quanta producing distinct frequencies of light in the visible region of the electromagnetic spectrum.

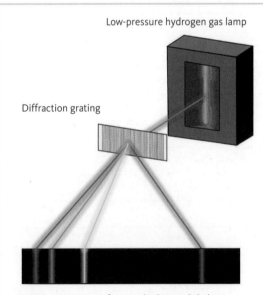

Low-pressure hydrogen gas lamp

Diffraction grating

Emission spectrum of atomic hydrogen (H) shows only those frequencies emitted by the low-pressure gas lamp.

Optical absorption spectra

When white light from a distant source **shines through** a low-pressure gas, the resultant spectrum is continuous but with dark lines where certain frequencies of light are missing. These missing frequencies have been absorbed by the gas atoms because the energies of the absorbed photons **exactly match** the energies required to excite the gas atom by causing an electron to transition up to a higher level.

These excited gas atoms are unstable and re-radiate photons with the same energies, but in all directions, so the intensity in the direction of the screen is reduced for those frequencies, resulting in dark lines.

The dark lines in an absorption spectrum appear in the same positions as the coloured lines in the emission spectrum of the same gas.

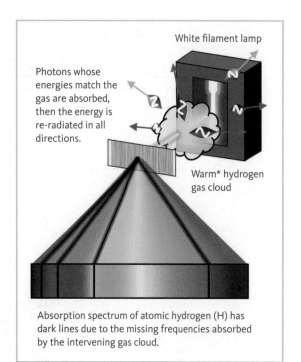

Photons whose energies match the gas are absorbed, then the energy is re-radiated in all directions.

White filament lamp

Warm* hydrogen gas cloud

Absorption spectrum of atomic hydrogen (H) has dark lines due to the missing frequencies absorbed by the intervening gas cloud.

* If the hydrogen gas cloud that is inbetween the light source and the diffraction grating is very cold, most of the electron transitions will be to the $n = 1$ energy level, resulting in mostly UV light being absorbed and re-radiated. This means that virtually all of the visible light photons will pass straight through without being absorbed, as they have insufficient energy to excite the electron up to the $n = 2$ level. Consequently, no dark lines will appear in the visible spectrum. However, if the gas is heated so that the electrons in the atoms are transitioning to and from the $n = 2$ energy level, then visible light photons will be absorbed, producing the absorption spectrum seen in the diagram above.

By studying the emission and absorption spectra of stars and nebulae (such as the Tarantula nebula, shown on the right), we are able to discover not only their composition, but also their temperature, age, speed of rotation and motion through space.

Rydberg's formula

JJ Balmer studied the visible spectra of monatomic hydrogen (**Balmer series**) and constructed a mathematical relationship between the wavelength of the light and the energy level from which the electron transitions down to the $n = 2$ level. Johannes Rydberg then developed the formula so that it could be used to accurately determine the wavelength of all electromagnetic waves produced by an electron transition between any higher energy level, L, and any lower energy level, S, in monatomic hydrogen. His formula was later verified by Theodore Lyman and Friedrich Paschen, who observed similar patterns in the ultraviolet (Lyman series) and infrared (Paschen series).

$$\frac{1}{\lambda} = R\left(\frac{1}{S^2} - \frac{1}{L^2}\right)$$

where

Rydberg's constant: $R = 1.097 \times 10^7 \text{ m}^{-1}$
series number: $S = 1$ for Lyman (ultraviolet), 2 for Balmer (visible), 3 for Paschen (infrared)
higher energy level: $L = S + \{1, 2, 3, 4, 5, 6, ..., \infty\}$

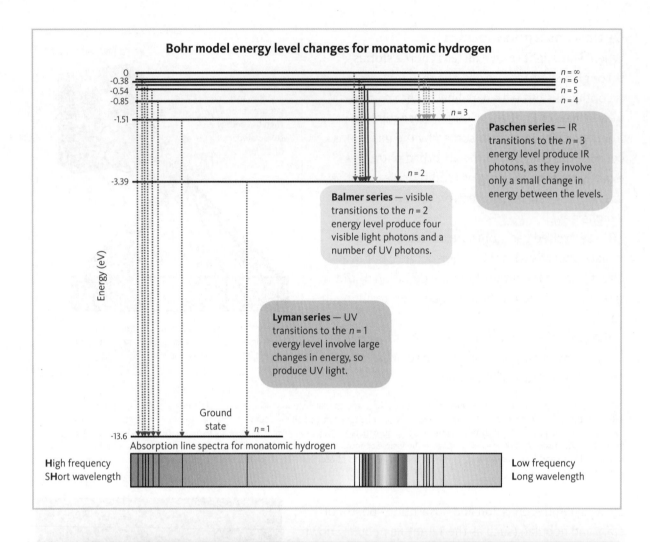

Bohr model energy level changes for monatomic hydrogen

Paschen series — IR transitions to the $n = 3$ energy level produce IR photons, as they involve only a small change in energy between the levels.

Balmer series — visible transitions to the $n = 2$ energy level produce four visible light photons and a number of UV photons.

Lyman series — UV transitions to the $n = 1$ evergy level involve large changes in energy, so produce UV light.

Ground state

Absorption line spectra for monatomic hydrogen

High frequency
SHort wavelength

Low frequency
Long wavelength

Worked example: Rydberg's formula

An infrared photon with a frequency of 2.727 x 10^{14} Hz is emitted as a result of a transition inside a monatomic hydrogen atom down to the $n = 3$ energy level.

With the aid of the diagram, determine the quantum number of the higher energy level from which the electron 'fell'.

Rydberg's constant $R = 1.097 \times 10^7$ m^{-1}

Speed of light $c = 2.998 \times 10^8$ m s^{-1}

-0.28 eV	$n = 7$
-0.38 eV	$n = 6$
-0.54 eV	$n = 5$
-0.85 eV	$n = 4$
-1.51 eV	$n = 3$

Solution

When using Rydberg's formula, $\dfrac{1}{\lambda} = R\left(\dfrac{1}{S^2} - \dfrac{1}{L^2}\right)$, it is helpful to remember that that the lower energy level is always the **S**maller number and the higher energy level is always the **L**arger number.

Given $\quad f = 2.727 \times 10^{14}$ Hz, $S = 3$, $R = 1.097 \times 10^7$ m⁻¹, $c = 2.998 \times 10^8$ m s⁻¹

Unknown $\quad L = ?$

Equations
$$v = f\lambda \text{ and } \frac{1}{\lambda} = R\left(\frac{1}{S^2} - \frac{1}{L^2}\right)$$

Substitute

$\lambda = \dfrac{2.998 \times 10^8}{2.727 \times 10^{14}} = 1.099 \times 10^{-6}$ m $\qquad$ *Substitute the wavelength into Rydberg's formula:*

$\dfrac{1}{1.099 \times 10^{-6}} = 1.097 \times 10^7 \left(\dfrac{1}{3^2} - \dfrac{1}{L^2}\right) \qquad$ *Dividing both sides by 1.099 x 10⁻⁶:*

Solve

$\dfrac{1}{(1.099 \times 10^{-6}) \times (1.097 \times 10^7)} = \dfrac{1}{9} - \dfrac{1}{L^2} \qquad$ *Solve the left-hand side and isolate the unknown:*

$\dfrac{1}{L^2} = \dfrac{1}{9} - \dfrac{1}{12.06} = 0.02817 \qquad$ *Find the square root:*

$\dfrac{1}{L} = 0.1678 \qquad$ *Find the reciprocal:*

$L = \dfrac{1}{0.1680} = 5.958$ *and as quantum numbers are integers,* $\mathbf{L = 6}$

Discrete atomic energy levels

By combining Rydberg's formula with the wave equation $v = f\lambda$ and Planck's equation $E = hf$, the energy of a transition can be calculated directly using:

$$E = hcR\left(\frac{1}{S^2} - \frac{1}{L^2}\right)$$

Expanding this formula gives:

$$E = hcR\frac{1}{S^2} - hcR\frac{1}{L^2}$$

which states:

$$\frac{\text{energy of the}}{\text{emitted photon}} = \frac{\text{energy of the lowest}}{\text{energy level}} - \frac{\text{energy of the}}{\text{excited energy}}$$

So the discrete energy E_n of any level n in a hydrogen atom can be found using the formula:

$$E_n = -hcR\,\frac{1}{n^2}$$

Extension concepts for scholarship — Spontaneous and stimulated emission

When an atom absorbs a photon of energy E with exactly the right amount of energy, it causes the electron to transition from a lower energy level, E_1, to a higher energy level, E_2, in a process called **stimulated absorption**, where $E = E_2 - E_1$.

The excited atom is unstable and the electron will eventually transition back down to the lower energy level, emitting a photon of energy $E = E_2 - E_1$ in a process called **spontaneous emission**.

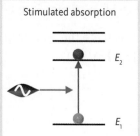

Stimulated absorption

However, if an atom that has already been excited by a photon of energy E is struck by another photon of the **same energy**, the electron can be forced to transition back to the lower level in a process known as **stimulated emission**. Not only do the incident photon and the emitted photon have the same frequency, but they move in the same direction and have the same phase (coherent).

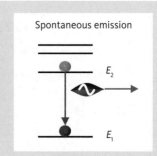

Spontaneous emission

Normally, the number of unexcited atoms in a gas is greater than the number of excited atoms. But by 'pumping' energy into a gas, the number of excited atoms in the gas can be made to exceed the number of unexcited atoms (a situation referred to as a population inversion). Now the chance of stimulated emission occurring increases, and the additional photon released increases the chance of further stimulated emissions. A chain reaction occurs producing **L**ight with an **A**mplification of intensity due to the **S**timulated **E**mission of a large number of identical photons all travelling together as a coherent beam of **R**adiation — the operating principles of a LASER.

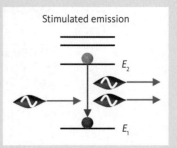

Stimulated emission

Photons produced by stimulated emission inside a laser are incident on two mirrors at either end that cause them to bounce back and forth triggering the stimulated emission of even more photons. One of the mirrors is partially transparent and about 1% of the photons pass through to form the laser beam.

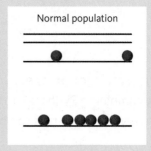

Normal population

Population inversion

The narrow high-intensity beam of lasers have a huge number of applications such as architecture, astronomy, communications, dentistry, engineering, entertainment, industry, holography, medicine, military and optometry.

Exercise 5E

You may need to use the following constants in the questions below.

Planck's constant: $h = 6.626 \times 10^{-34}$ J s
Rydberg's constant: $R = 1.097 \times 10^7$ m^{-1}
Speed of light: $c = 2.998 \times 10^8$ m s^{-1}
eV to joule conversion: 1 eV = 1.60×10^{-19} J

1 Calculate the energy of the seventh energy level of monatomic hydrogen. Express your answer in J and eV. (A)

_____

_____

_____

   ISBN: 9780170368179

2   Calculate the quantum number of the energy level in monatomic hydrogen with an energy of −0.1688 eV. (M)

_____

_____

_____

3   An electron falls from the eighth energy level of monatomic hydrogen emitting a photon of wavelength 387.9 nm. Determine the shell to which the electron falls and name the series. (M)

_____

_____

_____

_____

_____

_____

4   When a hydrogen discharge tube is heated, it glows a pink-purple colour. When the light is viewed through a spectroscope, only four distinct coloured lines can be seen: red, blue-green, violet, and deep violet. Discuss why only these colours are observed. (E)

_____

_____

_____

_____

_____

_____

_____

5   White light from the Sun travels through a low-pressure gas cloud. A photon is absorbed by monatomic hydrogen in the gas causing an electron to transition from the $n = 1$ to the $n = 4$ energy level. The electron spontaneously falls back down to the $n = 1$ energy level, re-emitting light of the same colour. When photographed using a spectroscope, a dark line is observed in the spectrum.

a   State the colour of the light missing from the continuous spectrum and the name of the series in which the dark line appears. (A)

_____

_____

_____

b   Explain why a dark line appears even though the photon is re-emitted. (M)

_____

_____

_____

_____

c Calculate the wavelength of the emitted photon. (A)

_____

_____

_____

d Determine the energy of the emitted photon and express the answer in eV. (M)

_____

_____

_____

_____

_____

6 A UV photon is absorbed by a low-pressure monatomic hydrogen gas cloud causing the electron to be promoted from the ground state to the $n = 6$ energy level. The electron falls back to the ground state in two steps releasing a UV photon of energy 12.155 eV and an infrared photon.

a Calculate the wavelength of the infrared photon. (M)

_____

_____

_____

_____

_____

_____

_____

b Describe all alternative transition routes that would result in a visible photon being produced. (A)

_____

_____

_____

_____

_____

_____

7 Explain why a solid produces a continuous spectrum but a low-pressure gas produces an emission spectrum. (M)

_____

_____

_____

_____

_____

ISBN: 9780170368179

Scholarship question

8   When hydrogen gas clouds in deep space are heated by a nearby star, they begin to glow a pink colour. Spectral examination of the light from the clouds reveals bright lines that shows that they are made up of hydrogen. However, when white light from a star whose light passes through the cloud is observed, it reveals dark lines in the same positions as the bright lines. Discuss why these findings are not contradictory and state any assumptions you have made.

_____

_____

_____

_____

_____

_____

_____

_____

_____

_____

5.3 Nuclear physics

Nuclear forces, mass and energy

The strong interaction and electrostatic repulsion

If the electrostatic (or Coulombic) force was the only force acting inside a nucleus, then the positive charge on the protons would cause them to fly apart due to electrostatic repulsion and the atom would collapse. Once outside the nucleus, a free neutron is unstable and decays with a half-life of about 11 minutes, but when combined in a nucleus with protons and other neutrons, it becomes stable. So what holds the nucleus together and stabilises the neutrons?

The nucleus of an atom is held together by an attractive force called the **strong interaction** or strong force that acts between all the protons and neutrons and is strong enough to overcome the electrostatic repulsion that is pushing the protons apart and prevents the decay of the neutrons.

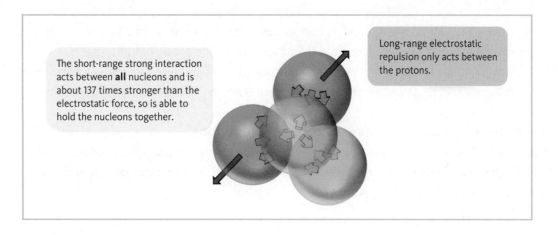

The short-range strong interaction acts between **all** nucleons and is about 137 times stronger than the electrostatic force, so is able to hold the nucleons together.

Long-range electrostatic repulsion only acts between the protons.

The strong interaction is **very short range** with a limit of about 1×10^{-15} m (that is, the diameter of the nucleus) and it does not affect electrons. Small atoms tend to be very stable due to the close proximity of all the nucleons to each other. But in larger atoms the nucleons on either edge may be so far apart that the strong interaction does not hold them as strongly.

Conservation conundrums

Physicists are able to determine the outcome of a nuclear reaction by using the following principles of conservation:

- charge
- momentum
- mass number, and
- atomic number.

However, two of the most fundamental principles of conservation appear to fail during nuclear reactions:

- Mass can be created or destroyed, in contradiction of the principle of conservation of mass.
- Energy before and after a nuclear reaction doesn't remain constant, invalidating the principle of conservation of energy.

Einstein's solution

In 1905, Einstein solved the problem by suggesting that energy has mass and that any reaction that involves a change in energy must also involve an equivalent change in mass. This means that when mass is transferred, an equivalent amount of energy is also transferred.

Einstein derived the following relationship:

energy change (J) = change in mass (kg) x speed of light^2 (m^2 s^{-2})

Expressed mathematically:

$$\Delta E = \Delta mc^2$$

Rest mass, m_0

The **rest mass**, m_0, of an object is the mass as measured when the object is at rest relative to the observer. An object with rest mass will also have **rest energy**, $E_0 = m_0c^2$.

A moving object has kinetic energy, and as energy has mass, the mass of an object must increase as its speed increases. At 'slow' speeds (less than about 1% of the speed of light; that is, 3 000 000 m s^{-1}), this additional mass is negligible, but as an object approaches 5% of the speed of light, the increase in mass becomes measurable and starts to affect experimental results.

Chemical and physical reactions versus nuclear reactions

For everyday chemical and physical changes in energy, the change in mass is extremely small and can be ignored. For example, when 1 kg of water cools by 1 °C it releases energy and as a consequence its mass decreases by 4.67×10^{-14} kg.

However, nuclear reactions, such as the radioactive decay of an element, involve much more significant changes in mass and therefore the energy released is much greater. For example, when 1 kg of radium decays into radon, the total mass decreases by 2.314×10^{-5} kg, resulting in an equivalent decrease in the energy of the atoms. This energy is released as heat, kinetic and radiant (for example gamma rays) energy.

This change in mass is 495 million times greater than the change in mass for the cooling water example above. Consequently, the amount of energy released by nuclear reactions is significantly greater than the energy released during chemical and physical reactions.

The principle of conservation of mass-energy

When dealing with nuclear reactions, physicists must consider the total mass-energy before and after a nuclear reaction rather than mass and energy as separate quantities.

> The total mass-energy before a reaction equals the total mass-energy afterwards.

which can be stated as follows:

| rest energy of the reactants | + | kinetic/radiant energy of the reactants | = | rest energy of the products | + | kinetic/radiant energy of the products |

Expressed mathematically:

$$m_{reactants}c^2 + E_{reactants} = m_{products}c^2 + E_{products}$$

During spontaneous decay, the reactant changes (transmutes) into the daughter products without any energy being supplied. When the rest mass of the products is less than the rest mass of the reactant, energy is released. If the rest mass of the products is greater than the reactants, then energy must be supplied to cause the decay to occur. This is called artificial transmutation.

Worked example: Artificial transmutation by Rutherford

In 1919, Rutherford bombarded nitrogen gas with alpha particles resulting in the artificial transmutation of nitrogen into oxygen. The reaction is shown below.

$$^{14}_{7}N + ^{4}_{2}He \rightarrow ^{16}_{8}O + ^{1}_{1}H$$

Particle	Mass
Nitrogen	23.24627×10^{-27} kg
Helium	6.64466×10^{-27} kg
Oxygen	28.22043×10^{-27} kg
Hydrogen	1.67262×10^{-27} kg

After the reaction, the oxygen and hydrogen move apart with a kinetic energy of 1.88170×10^{-13} J shared between them. Speed of light $c = 3.00 \times 10^8$ m s^{-1}.

Determine the velocity of the incident alpha particle required to cause this transmutation, assuming that the nitrogen nucleus was at rest when it was hit by the alpha particle.

Solution

First use the principle of conservation of mass-energy to find the energy associated with the reactants, then use this to determine the kinetic energy of the alpha particle.

Given

Reactants: $m_N = 23.24627 \times 10^{-27}$ kg and $m_\alpha = 6.64466 \times 10^{-27}$ kg

Products: $m_O = 28.22043 \times 10^{-27}$ kg and $m_H = 1.67262 \times 10^{-27}$ kg

Products: $E_K = 1.88170 \times 10^{-13}$ J

Speed of light: $c = 3.00 \times 10^8$ m s^{-1}

Unknown

$v_\alpha = ?$

Equations

$$m_{reactants}c^2 + E_{reactants} = m_{products}c^2 + E_{products}$$

$$E_K = \frac{1}{2}mv^2$$

Substitute

$m_{reactants} = 23.24627 \times 10^{-27} + 6.64466 \times 10^{-27} = 29.89093 \times 10^{-27}$ kg

$m_{products} = 28.22043 \times 10^{-27} + 1.67262 \times 10^{-27} = 29.89305 \times 10^{-27}$ kg

Mass energy$_{reactants} = 29.89093 \times 10^{-27} \times (3.00 \times 10^8)^2 + E_{reactants}$

Mass energy$_{products} = 29.89305 \times 10^{-27} \times (3.00 \times 10^8)^2 + 1.88170 \times 10^{-13}$

Equating: mass energy$_{reactants}$ = mass energy$_{products}$

$2.6901837 \times 10^{-9} + E_{reactants} = 2.6903745 \times 10^{-9} + 1.88170 \times 10^{-13}$

$E_{reactants} = 2.69056267 \times 10^{-9} - 2.6901837 \times 10^{-9} = 3.7897 \times 10^{-13}$ J

The additional energy of the reactants is from the E_K of the alpha particle,

so $E_{reactants} = \frac{1}{2}mv^2$ hence $3.7897 \times 10^{-13} = \frac{1}{2} \times 6.64466 \times 10^{-27} \times v^2$

Solve

$$v = \sqrt{\frac{3.7897 \times 10^{-13}}{3.32233 \times 10^{-27}}} = 1.0680 \times 10^7 = 1.07 \times 10^7 \text{ m s}^{-1} \text{ (3 sf)}$$

Exercise 5F

Where required, use the following constants: speed of light $c = 2.998 \times 10^8$ m s^{-1}; electron volt, 1 eV $= 1.602 \times 10^{-19}$ J.

1 An aluminium pan used for cooking decreases in mass by 8.09×10^{-13} kg when it cools back down to room temperature. Calculate the energy released during cooling. (A)

_____

_____

_____

2   Calculate the change in mass of an electric jug filled with 1.5 kg of water when 504 kJ of energy is supplied to boil the water. State any assumptions that you make in determining your answer. (Hint: think about what happens when you boil a real kettle.) (A)

_____

_____

_____

_____

_____

3   Explain why the mass of an ice cube won't be exactly the same as the mass of the water from which it was formed. (M)

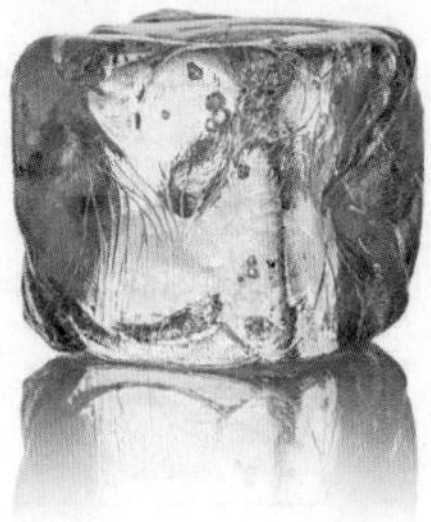

_____

_____

_____

_____

4   Why can scientists use the principle of conservation of mass in their calculations of everyday chemical and physical changes without fear of getting the wrong answer? Use your answers to questions **1** and **2** to justify your answer. (M)

_____

_____

_____

_____

_____

5   State the principles of conservation that scientists can use to solve nuclear reaction problems. (A)

_____

_____

_____

6   Describe the strong force and explain why it is easier to break large atoms apart compared with small atoms. (M)

_____

_____

_____

_____

_____

_____

_____

7   Uranium 238 is often used for dating rock deposits, as it has an extremely long half-life. It decays to thorium 234 by releasing an alpha particle in the process.

    a   Complete the nuclear equation below and explain how you were able to determine the values. (A)

$$^{238}_{92}U \longrightarrow {}^{234}_{90}Th + {}^{\square}_{\square}\alpha + energy$$

_____

_____

_____

    b   Using the values in the table opposite, calculate the energy released by a stationary uranium atom during this spontaneous nuclear decay. (M)

Particle	Mass
Uranium 238	395.209×10^{-27} kg
Thorium 234	388.557×10^{-27} kg
Alpha particle	6.645×10^{-27} kg

_____

_____

_____

_____

_____

_____

    c   Explain why the thorium nucleus moves in the opposite direction to the alpha particle at a much slower speed. (E)

_____

_____

_____

_____

_____

_____

Scholarship question

8   An americium 241 nucleus is at rest when it spontaneously decays into neptunium 237 and an alpha particle. The two products fly apart in opposite directions. Determine the percentage of the kinetic energy taken by the emitted alpha particle during the decay.

_____

_____

_____

_____

_____

   ISBN: 9780170368179

_____

_____

_____

_____

_____

_____

_____

_____

_____

_____

_____

_____

_____

_____

_____

Mass defect and binding energy

Mass defect, Δm

Experiments reveal that the mass of any nucleus (except hydrogen, which contains only one proton) is less than the sum of the masses of the individual nucleons (protons plus neutrons). This difference in the mass is known as the **mass defect**, Δm.

mass defect (kg) = mass of the **nucleons** (kg) – mass of the **nucleus** (kg)

Expressed mathematically:

$$\Delta m = m_{nucleons} - m_{nucleus}$$

Binding energy, E

The nucleons in a nucleus are tightly bound together by the strong interaction and work must be done to separate the nucleons and set them free from the nucleus.

The total work done (energy supplied) to break a nucleus into separate nucleons is called the **binding energy**.

As energy must be added to the nucleons to break them free of the nucleus, the free nucleons must have more energy and hence more mass than nucleons which are bound together in the nucleus. The binding energy of a nucleus is the energy equivalent of the mass defect and can be calculated using Einstein's mass-energy relation $\Delta E = \Delta mc^2$:

binding energy (J) = mass defect (kg) x speed of light^2 (m^2 s^{-2})

Binding energy per nucleon

The stability of a nucleus is described by the average amount of energy required to break each individual nucleon free and is referred to as the binding energy per nucleon (BEPN).

$$\text{binding energy per nucleon (J)} = \frac{\text{total binding energy (J)}}{\text{number of nucleons}}$$

Expressed mathematically:

$$BEPN = \frac{\Delta E}{A}$$

Worked example: The mass defect and binding energy of helium

A $^{4}_{2}$He nucleus has a mass of 6.64466×10^{-27} kg and is made up of two protons and two neutrons. Using the values given opposite, determine:

Particle	Mass
Neutron	1.67493×10^{-27} kg
Proton	1.67262×10^{-27} kg

a the mass defect for helium

b the total binding energy required to completely separate a helium nucleus into its separate parts, and

c the binding energy per nucleon in MeV.

Speed of light $c = 3.00 \times 10^{8}$ m s^{-1}

Solution

a First find the mass of the parts of the nucleus (the nucleons) and compare it with the mass of the nucleus.

Mass of neutrons, $^{1}_{0}n$

$2 \times 1.67493 \times 10^{-27}$

$= 3.34986 \times 10^{-27}$ kg

Mass of protons, $^{1}_{1}p$

$2 \times 1.67262 \times 10^{-27}$

$= 3.34524 \times 10^{-27}$ kg

Mass of the nucleus, $^{4}_{2}$He^{2+}

$m_{nucleus} = 6.64466 \times 10^{-27}$ kg

Mass of the nucleons (neutrons plus protons)

$m_{nucleons} = 3.34986 \times 10^{-27} + 3.34524 \times 10^{-27}$

$m_{nucleons} = 6.69510 \times 10^{-27}$ kg

    ISBN: 9780170368179

Now find the mass defect by calculating the difference.

Given	$m_{nucleons} = 6.69510 \times 10^{-27}$ kg, $m_{nucleus} = 6.64466 \times 10^{-27}$ kg
Unknown	$\Delta m = ?$
Equations	$\Delta m = m_{nucleons} - m_{nucleus}$
Substitute	$\Delta m = 6.69510 \times 10^{-27} - 6.64466 \times 10^{-27}$
Solve	$\Delta m = 0.05044 \times 10^{-27}$ kg

b The total binding energy of helium is given by Einstein's mass-energy relation.

Given	$\Delta m = 0.05044 \times 10^{-27}$ kg, $c = 3.00 \times 10^8$ m s^{-1}
Unknown	$\Delta E = ?$
Equations	$\Delta E = \Delta mc^2$
Substitute	$\Delta E = 0.05044 \times 10^{-27} \times (3.00 \times 10^8)^2$
Solve	$\Delta E = 4.54 \times 10^{-12}$ J

This means that 4.54×10^{-12} J of energy is required to completely separate all the nucleons in the helium nucleus.

c The stability of the nucleus is indicated by the binding energy per nucleon.

Given	$\Delta E = 4.54 \times 10^{-12}$ J, number of nucleons $A = 4$
Unknown	$BEPN = ?$
Equations	$BEPN = \dfrac{\Delta E}{A}$
Substitute	$BEPN = \dfrac{4.54 \times 10^{-12}}{4}$
Solve	$BEPN = 1.1349 \times 10^{-12}$ J
	So the binding energy per nucleon in MeV is given by:
	$BEPN = \dfrac{1.1349 \times 10^{-12}}{1.6 \times 10^{-19}} = 7.09 \times 10^6 = 7.09$ MeV

Nuclear reactions

Binding energy and stability

The greater the binding energy per nucleon, the more stable the nucleus becomes as more energy is required to remove each nucleon. The graph below shows how the binding energy per nucleon varies with nucleon number.

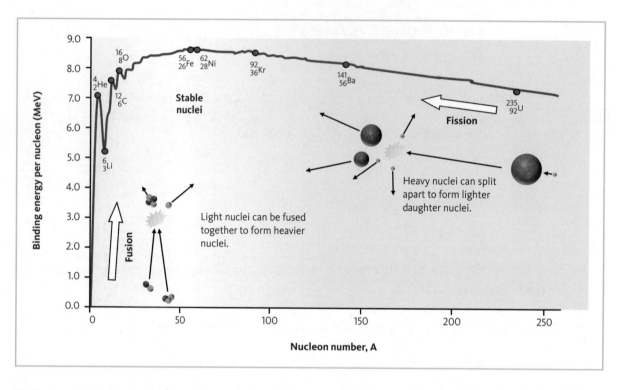

Hydrogen 1 has no binding energy, as it contains only an unbound proton. The binding energy per nucleon then rises rapidly in a series of peaks up to iron and nickel, which are found at the top of the curve and are the most stable nuclei with a binding energy per nucleon of about 8.8 MeV. The binding energy per nucleon then gradually decreases as the nucleon number increases and the nuclei become less stable.

Nuclear potential energy

A graph of nuclear potential energy versus nucleon number is the inverse of the binding energy per nucleon graph and is a useful tool for considering the energy changes during fission and fusion reactions.

The graph shows how much work must be done, per nucleon, to overcome the strong force and remove a nucleon from a nucleus to a point where it will have a potential energy of $E_p = 0$ MeV.

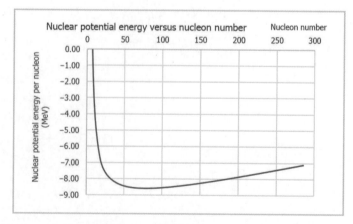

As energy must be supplied to bring the nucleon to $E_p = 0$ MeV, it means that the bound nucleon must have had less than $E_p = 0$ MeV when bound, that is, a negative nucleon potential energy. The amount of energy required to unbind the nucleon is equal to the binding energy per nucleon.

Gravitational analogy

Consider ball **1** at rest on the surface where it has no gravitational potential energy. Next to ball **1** is a deep ('potential') well in which any ball will have less gravitational potential energy because it is lower down. Ball **2** is at rest at the deepest part of the well. To remove the ball from the well will require the greatest amount of work to be done on the ball to bring it up to $E_{gp} = 0$ J.

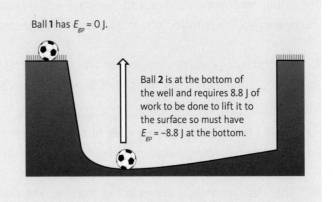

Ball 1 has $E_{gp} = 0$ J.

Ball **2** is at the bottom of the well and requires 8.8 J of work to be done to lift it to the surface so must have $E_{gp} = -8.8$ J at the bottom.

Fission

During fission reactions, heavy nuclei **split** to form lighter nuclei, which have a greater binding energy per nucleon and hence are more stable. During the reaction the total mass decreases, so the mass of the products is less than the mass of the reactants. As the products have a lower mass, they have a lower energy. This decrease in energy means that the products are more stable and have a greater binding energy per nucleon. The energy released by the reaction appears as E_K of the decay products, so the reaction is a source of heat.

Gravitational analogy

Ball **3** at the top of the slope is closer to the surface, so less work must be done to remove it from the well.

But ball **3** is unstable, as it sits high up on the slope, and can roll down changing potential energy into kinetic energy until it reaches the bottom.

At **4**, the ball has less potential energy, so

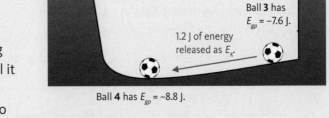

Ball 3 has $E_{gp} = -7.6$ J.

1.2 J of energy released as E_K.

Ball 4 has $E_{gp} = -8.8$ J.

requires more work to be done to remove it from the well so is more stable.

Fusion

During fusion reactions, light nuclei are fused together to form heavier nuclei, which have a greater binding energy per nucleon and so are more stable. For fusion reactions to occur, the nuclei must be moving extremely fast. A lot of kinetic energy is require to do work against the electrostatic force to get the nuclei close enough so that the strong force can take effect. Consequently, fusion reactions require extremely high temperatures (several million degrees Celsius) to start.

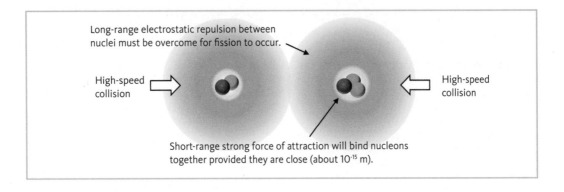

Long-range electrostatic repulsion between nuclei must be overcome for fission to occur.

High-speed collision

High-speed collision

Short-range strong force of attraction will bind nucleons together provided they are close (about 10^{-15} m).

The initiating energy is not part of the mass-energy exchange in the fusion reaction. The **total mass decreases** and the products have a lower energy, increasing their stability and releasing energy.

Conservation of mass-energy for nuclear reactions

Both fission and fusion reactions release energy due to a difference in the mass of the reactants and products.

- Fusion releases more energy **per nucleon** than fission.
- There are far fewer nucleons in light nuclei than in the heavier nuclei involved with fission, so fission releases more energy **per atom**.
- There are more light nuclei in 1 kg than heavier nuclei, so fusion releases more energy **per kilogram**.

The energy released can be calculated using the principle of conservation of mass-energy.

Exercise 5G

Where required, use the following constants: $c = 2.998 \times 10^8$ m s^{-1}, 1 eV = 1.602×10^{-19} J.

1 Nickel 62 is the most stable nuclei due to its binding energy and the compact nature of its nucleus. A $^{62}_{28}$Ni nucleus has a mass of $102.80892 \times 10^{-27}$ kg. Using the data given in the table, determine the following values.

Particle	Mass
Neutron	1.67493×10^{-27} kg
Proton	1.67262×10^{-27} kg

a The number of protons and neutrons found in an atom of nickel 62. (A)

_____

_____

b The mass defect for nickel 62. (M)

_____

_____

_____

c The total binding energy in MeV required to completely separate a nickel nucleus into its separate parts. (A)

_____

_____

_____

d The binding energy per nucleon of nickel 62 in MeV. (A)

_____

_____

_____

2   Lithium, $^6_3$Li, is a stable nucleus often used in nuclear fusion reactions due to its low binding energy per nucleon of 5.332621 MeV. Using the data given in the table, determine the following values.

Particle	Mass
Neutron	1.67493×10^{-27} kg
Proton	1.67262×10^{-27} kg
Electron	9.10940×10^{-31} kg

a   The total binding energy in joules required to completely separate a lithium nucleus into its separate parts. (A)

_____

_____

b   The mass defect of a lithium nucleus. (M)

_____

_____

_____

c   The mass of a neutral lithium 6 **atom**. (A)

_____

_____

_____

_____

3   The graph of binding energy per nucleon on page 240 spikes for helium, carbon and oxygen. By considering what these nuclei have in common, explain why heavier nuclei can decay by alpha emission, rather than by emitting random groups of neutrons and protons. (E)

_____

_____

_____

_____

4   The Sun's energy comes from a series of different fusion reactions involving isotopes of hydrogen. During one of the fusion reactions, hydrogen and deuterium fuse together to form helium 3.

Particle	Mass
Neutron	1.67493×10^{-27} kg
Hydrogen	1.67262×10^{-27} kg
Deuterium	3.34358×10^{-27} kg
Helium 3	5.00641×10^{-27} kg

$$^1_1H + {}^2_1H \rightarrow {}^3_2He + {}^0_0\gamma$$

a   Show that 5.493 MeV of energy is released by the reaction as a gamma photon. (E)

_____

_____

_____

_____

_____

_____

b Calculate the maximum frequency of the gamma photon and state any assumptions you are making. (Planck's constant $h = 6.626 \times 10^{-34}$ J s.) (M)

_____

_____

_____

_____

Other than hydrogen 1, all the elements in the universe up to iron and nickel have been formed during the fusion processes inside stars as they release energy. Elements like uranium and plutonium, which are heavier then iron and nickel, are formed through fusion during supernovae — a massive explosion that occurs at the death of a supergiant star.

c Iron 58 has a higher binding energy per nucleon than uranium 238, but uranium has a much greater binding energy. Discuss this statement and explain which is more stable. (E)

_____

_____

_____

_____

_____

_____

d By considering the graph on page 240, discuss why the formation of elements with a greater mass than iron and nickel only occurs inside supernovae. (E)

_____

_____

_____

_____

_____

_____

_____

_____

_____

_____

_____

_____

_____

 ISBN: 9780170368179

Fusion releases far more energy than fission for the same mass of fuel, however nuclear power plants rely on the fission of uranium or plutonium to produce electricity.

e Show that the energy released during a nuclear fission reaction is 179.64 MeV. (E)

$${}^{1}_{0}n + {}^{235}_{92}U \rightarrow {}^{144}_{56}Ba + {}^{90}_{36}Kr + 2{}^{1}_{0}n + energy$$

Particle	Mass
Neutron	1.67493×10^{-27} kg
Uranium 235	390.216×10^{-27} kg
Barium 144	238.939×10^{-27} kg
Krypton 90	149.282×10^{-27} kg

f With reference to the graph on page 240, explain why energy is released during this reaction. (M)

g By comparing your answers to questions **4a** and **4e**, explain why fusion releases 'more energy for the same mass of fuel' compared with fission. (M)

5 Use the values opposite to determine if a stationary oxygen 16 nucleus could spontaneously decay into a carbon 12 nucleus releasing an alpha particle.
The proposed nuclear decay is shown below. (E)

$$^{16}_{8}O \rightarrow ^{12}_{6}C + ^{4}_{2}\alpha$$

Particle	Mass
Oxygen 16	26.553×10^{-27} kg
Carbon 12	19.976×10^{-27} kg
Alpha particle	6.645×10^{-27} kg

_____

_____

_____

_____

_____

_____

_____

_____

_____

_____

6

Electrical systems

Achievement Standard 91526 (P3.6) requires students to demonstrate understanding by connecting concepts or principles that relate to electrical systems. This standard is worth 6 credits and is assessed externally.

DC circuits

- Internal resistance.
- Parallel plate capacitor, capacitance, dielectrics.
- Energy stored in a capacitor.
- Q/t, V/t and I/t graphs for a capacitor, time constant.
- V/t and I/t graphs for an inductor, time constant.
- Faraday's Law.
- The inductor.
- Energy stored in an inductor.

- Simple application of Kirchhoff's laws.
- Series and parallel capacitors.
- Magnetic flux, magnetic flux density.

- Lenz's Law.
- Self-inductance.
- The transformer.

AC circuits

- Comparison of the energy dissipation in a resistor carrying direct current and alternating current.
- Peak and rms voltage and current.
- Voltage and current and their phase relationship in LR and CR series circuits.
- Phasor diagrams.
- Reactance and impedance and their frequency dependence in a series circuit.
- Resonance in LCR circuits.

Relationships

$V = Ed$	$\Delta E = qV$	$E = \frac{1}{2}QV$	$Q = CV$
$C = \frac{\varepsilon_o \varepsilon_r A}{d}$	$C_T = C_1 + C_2 + C_3 \ldots$	$\frac{1}{C_T} = \frac{1}{C_1} + \frac{1}{C_2} + \frac{1}{C_3} \ldots$	$\tau = RC$
$R_T = R_1 + R_2 + R_3 \ldots$	$\frac{1}{R_T} = \frac{1}{R_1} + \frac{1}{R_2} + \frac{1}{R_3} \ldots$	$V = IR$	$P = IV$
$\phi = BA$	$\varepsilon = -L\frac{\Delta I}{\Delta t}$	$\varepsilon = -\frac{\Delta \phi}{\Delta t}$	$\frac{N_P}{N_S} = \frac{V_P}{V_S}$
$E = \frac{1}{2}LI^2$	$\tau = \frac{L}{R}$	$I = I_{max} \sin \omega t$	$V = V_{max} \sin \omega t$
$I_{max} = \sqrt{2}I_{rms}$	$V_{max} = \sqrt{2}V_{rms}$	$X_C = \frac{1}{\omega C}$	$X_L = \omega L$
$V = IZ$	$\omega = 2\pi f$	$f = \frac{1}{T}$	$f_0 = \frac{1}{2\pi \sqrt{LC}}$

6.0 Electric circuits basics

Quantities

Charge (Q)

Charge is a fundamental property of matter and occurs in two forms described as positive and negative. The quantity of charge, Q, on an object is measured in coulombs (C). The charge on electrons and protons is extremely small:

Charge on an electron, $q_e = -1.60 \times 10^{-19}$ C
Charge on a proton, $q_p = +1.60 \times 10^{-19}$ C

Current (I)

Throughout the book, the term current will refer to the convention of positive charge flow. Where required, the motion of electrons will be referred to as electron flow.

Current is defined as the flow of positive charges around a circuit each second. If a quantity of charge, Q, passes a given point in time, t, then a steady current, I, flows, hence:

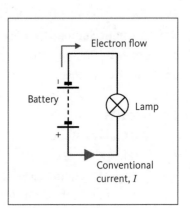

$$\text{current (A)} = \frac{\text{charge (C)}}{\text{time (s)}}$$

Expressed mathematically:

$$I = \frac{Q}{t}$$

Electric force (F)

When a complete circuit involving wires and a load is connected to the terminals of a battery, an electric field forms almost instantly inside the circuit, exerting a force on all the charge carriers in the circuit and causing them to accelerate. Each charge (q) in the field experiences a force (F), the strength of which can be determined using the equation:

$$\text{force (N)} = \text{electric field strength (N C}^{-1}\text{)} \times \text{charge (C)}$$

Expressed mathematically:

$$F = Eq$$

When the electric force (F) causes the charge carriers to move a distance (d), work (W) is done, and energy is changed (ΔE), and this is given by the equation $\Delta E = W = Fd$.

Substituting in the electric force equation $F = Eq$ gives:

$$\Delta E = Eqd$$

which reveals that:

$$\text{change in energy (J)} = \text{electric field strength (V m}^{-1}\text{)} \times \text{charge (C)} \times \text{distance (m)}$$

 ISBN: 9780170368179

The charge carriers in the electric field have the potential to do work due to their position in the electric field (similar to a mass placed in a gravitational field), but only if it is able to move to a position of lower potential energy.

Voltage (V)

A **voltage** drop describes the amount of work done by each unit of charge (Q) as it moves through the load and can be represented by the following relationship:

$$\text{voltage (V)} = \frac{\text{energy changed from electrical potential (J)}}{\text{quantity of charge (C)}}$$

Expressed mathematically:

$$V = \frac{\Delta E}{q}$$

Combining with the electrical potential energy equation above gives $V = \dfrac{Eqd}{q}$.
Hence:

$$V = Ed$$

which states that:

$$\text{voltage (V)} = \text{electric field strength (V m}^{-1}\text{)} \times \text{distance (m)}$$

Extension concepts for scholarship — Drift velocity

The electric force causes the electrons to accelerate around the circuit transforming electrical potential energy into kinetic energy. But as the electrons collide with the atoms and ions inside the wires and components, they transfer kinetic energy to the atoms and ions causing them to vibrate more.

The force from the electric field causes the electrons to immediately accelerate again, and as a consequence of losing energy from collisions and gaining it from the electric field, the electrons move around the circuit with an average drift velocity less than a millimetre each second.

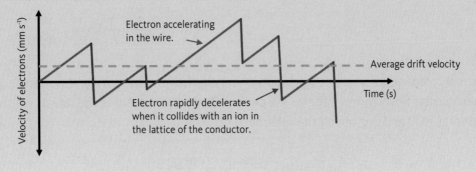

Conduction and resistance (R)

Collisions between the moving electrons and the atoms and ions inside the wires and components results in them becoming warmer and vibrating more. This increases the chance of collisions, which means the resistance of the conductor increases. If the temperature of a conductor is maintained at a constant value, then the relationship between voltage (V), current (I) and resistance (R) is given by:

voltage (V) = current (A) x resistance (Ω)

Expressed mathematically:

$$V = IR$$

Ohm's Law

Resistance in series and parallel

When resistors are connected in series, the **same current**, *I*, passes through each of them in turn. The total voltage across resistors in series is equal to the **sum of the individual voltages**.

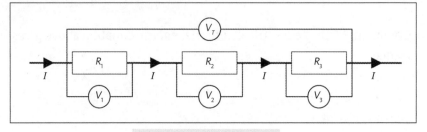

The total resistance due to resistors connected in series is equal to the sum of the individual resistances.

$$R_T = R_1 + R_2 + R_3$$

When resistors are connected in parallel, each branch will experience the **same potential difference, *V*.** The total current through resistors in parallel is equal to the **sum of the individual currents**.

With each additional parallel branch, the total resistance decreases due to the extra paths the current can take, so the total resistance is calculated using the equation shown at right.

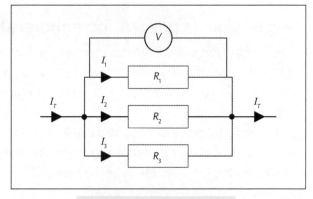

$$\frac{1}{R_T} = \frac{1}{R_1} + \frac{1}{R_2} + \frac{1}{R_3}$$

Power (P)

Some electrical components dissipate (give out) energy in the form of heat, light, sound and/or motion when current flows through them. The power (P) of a component is the rate at which it dissipates energy (ΔE) and is given by the relationship:

$$\text{power (W)} = \frac{\text{change in energy (J)}}{\text{time (s)}}$$

Expressed mathematically:

$$P = \frac{\Delta E}{t}$$

As $\Delta E = VQ$ and $I = \dfrac{Q}{t}$, the power (P) can also be expressed as follows:

power (W) = current (A) x voltage (V)

Expressed mathematically:

$$P = IV$$

Substituting Ohm's Law, $V = IR$, into the electrical power equation above provides two alternative power relationships in terms of the resistance (R) as follows:

as $P = IV$ and $V = IR$, then $P = I(IR)$, hence $\mathbf{P = I^2R}$

as $P = IV$ and $\dfrac{V}{R} = I$, then $P = \left(\dfrac{V}{R}\right)V$, hence $\mathbf{P = \dfrac{V^2}{R}}$

Open and closed circuits

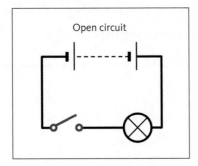

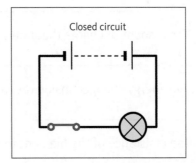

An open circuit is one in which no current can flow due to a break in the conducting path, by disconnecting a component, opening a switch or cutting the wires.

A closed circuit is one in which there is a complete path from the positive terminal of the power supply to the negative terminal, resulting in a flow of current.

Exercise 6A

1   Define current in terms of charge. (A)

_____

_____

2   Define voltage in terms of charge. (A)

_____

_____

_____

3 State Ohm's Law and any assumptions that must be considered when using it. (A)

_____

_____

4 Define power and show mathematically how power is related to resistance. (A)

_____

_____

_____

5 The lamp is rated at 6.0 V and 1.5 W and is connected to a battery as shown in the circuit opposite. Calculate the following quantities.

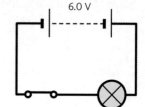

6.0 V

    a The energy supplied by the battery to each coulomb of charge. (A)

    _____

    b The current through the lamp. (A)

    _____

    c The amount of charge that flows through the lamp in 1 minute. (A)

    _____

    d The energy changed inside the lamp during 1 minute. (A)

    _____

    e The resistance of the filament in the lamp. (A)

    _____

    f The strength of the electric field in the filament if it is 30 cm long. (A)

    _____

6 Four resistors are connected in series as shown and connected to a 9.0 V power supply.

    a State the value of the current at _I_ and explain how you determined your answer.

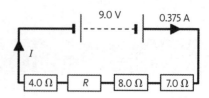

9.0 V   0.375 A

_I_

4.0 Ω   _R_   8.0 Ω   7.0 Ω

    _____

    _____

    _____

    b Show that resistor _R_ has a resistance of 5.0 Ω. (A)

    _____

    _____

    c Explain how you could determine the voltage across the 8.0 Ω resistor without using Ohm's Law ($V = IR$). (M)

    _____

    _____

d Calculate the rate at which heat energy is dissipated by the 8.0 Ω resistor. (A)

7 An engineer wishes to construct a number of electrical circuits but only has a 4.0 Ω resistor, a 20 Ω resistor and a 60 Ω resistor.

a Calculate the maximum resistance the engineer could achieve with these three resistors. (A)

b Calculate the minimum resistance the engineer could achieve with these three resistors. (A)

c How could the engineer achieve a resistance of 19 Ω with these three resistors. (M)

8 Explain the difference between an open and a closed circuit. (M)

9 Explain why a lamp in a circuit will light the instant the power is switched on despite the fact that electrons drift through the wires at about 0.0001 m s⁻¹. (M)

Scholarship questions

10 A standard torch lamp has a rating of 6 V and 1.5 W. However, as a battery in a torch runs down, it delivers less than the required 6 V to the lamp. Discuss what additional information is needed about the lamp to determine the behaviour of the torch when the battery is no longer operating at 6 V.

_____

_____

_____

_____

_____

_____

_____

_____

11 Inside a conducting wire of length l and cross-sectional area A, there are n free electrons per unit volume. Each electron has a charge e. When a potential is applied across the ends, the electrons drift through the wire at a velocity v. Prove that the current in the conductor is given by the relationship:

$$I = nAve$$

_____

_____

_____

_____

12 A nichrome wire of resistance R is bent into a circle. Show that the resistance between two points A and B, which are separated by an angle θ, is given by:

$$R_{AB} = R \left[\frac{2\pi\theta - \theta^2}{4\pi^2} \right]$$

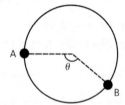

_____

_____

_____

_____

_____

_____

_____

    ISBN: 9780170368179

6.1 Electromotive force and internal resistance

Electromotive force (emf or ε)

The **electromotive force**, emf or ε (Greek letter epsilon), is defined as the energy **supplied** (ΔE) to each **unit of electrical charge** (Q) by an energy source such as a battery, electrical generator or solar cell.

$$\text{electromotive force (J C}^{-1}) = \frac{\text{energy supplied by the battery (J)}}{\text{quantity of charge (C)}}$$

Expressed mathematically:

$$\varepsilon = \frac{\Delta E}{Q}$$

Despite being called the electromotive *force*, the emf is not a force, so is not measured in newtons. Like the voltage drop across a resistor, emf is a **potential difference** and measures the change in electrical energy per unit charge, so is measured in volts, where **$1\,V = 1\,J\,C^{-1}$**.

However, emf and voltage drop describe different energy transfers in a circuit:

- emf (ε) refers to the conversion of 'supplied' energy into electrical energy.

- voltage drop (V) refers to the conversion of electrical energy into a 'load' energy.

Internal resistance

Real and ideal components

To simplify basic circuits, the components are considered to be ideal. **Ideal components** are 100% efficient, for example ideal ammeters have no electrical resistance and ideal voltmeters do not pass any current.

Real components are not 100% efficient, for example real ammeters have a small but measurable effect on the current that they are being used to measure and real voltmeters allow a small current to flow, reducing the total resistance in the circuit. This inefficiency is just as true for the power source.

Consider an **ideal** battery, where no energy is wasted and the amount of chemical energy converted by the chemical reactions is the same as the electrical potential energy provided to the external circuit.

However, in a **real** battery, some of the chemical energy is wasted as heat energy during the chemical reactions, so less electrical potential energy is available for the external circuit. This inefficiency can be modelled by considering a real battery as being made from an ideal battery in series with an internal resistance, r.

The circuit symbol for a real battery is shown in the diagram with a dotted box drawn round an ideal battery and an internal resistor to show that they represent a single component.

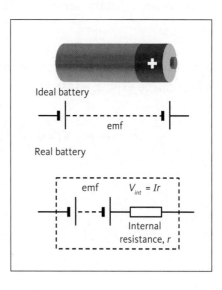

Terminal potential difference and emf
Open circuit

The terminal potential difference is measured by placing a voltmeter across the terminals (ends) of a real battery or power supply.

In an open circuit the terminal potential difference has its maximum value, which is equal to the emf of the battery.

In the example opposite, no current flows around the open circuit so there will be no voltage drop across the internal resistance of 3.75 Ω as $V = Ir$. $\qquad V_{internal} = 0 \times 3.75 = 0$ V

So the terminal potential will be: $\qquad V_{terminal} = \varepsilon = 9.0$ V

Closed circuit

When the switch is closed, a current now flows through the wires, external load (represented by the variable resistor), switch and battery. As a consequence:

In a closed circuit the terminal potential difference across the power supply decreases due to the internal resistance inside the battery.

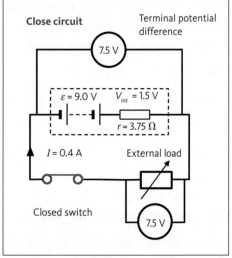

In the example opposite, the internal resistance of the battery is 3.75 Ω. When a current of 0.4 A flows there will be a voltage drop across the internal resistance: $V_{internal} = 0.4 \times 3.75 = 1.5$ V. This means that 1.5 V of the battery's emf has been wasted as thermal energy in the internal resistance and so only 7.5 V remains for the external load (variable resistor).

It can be seen that the terminal potential difference, $V_{terminal}$, is given by: $\qquad V_{terminal} = \varepsilon - V_{internal}.$

Substituting in Ohm's Law for the internal resistance as $V_{internal} = Ir$ gives: $\qquad \boxed{V_{terminal} = \varepsilon - Ir}$ which can be expressed as:

terminal pd (V) = emf of the battery (V) – current (A) x internal resistance (Ω)

 ISBN: 9780170368179

Worked example: Internal resistance and emf

The emf of a C type battery is 3.0 V. When the battery is connected across a lamp of resistance 5.0 Ω, a current of 0.45 A flows around the circuit.

a Determine the internal resistance, r, of the battery.

b Calculate the terminal potential of the battery when connected in the circuit.

$\varepsilon = 3.0$ V

$R = 5.0$ Ω

Solution

a A combined emf of 3.0 V is driving a current of 0.45 A through **BOTH** the internal and external resistances. The total resistance of the circuit will be:

Given	$\varepsilon = 3.0$ V, $I = 0.45$ A, $R_L = 5.0$ Ω
Unknown	$r = ?$
Equations	$\varepsilon = IR$ and $R_T = R + r$
Substitute	$R_T = \dfrac{\varepsilon}{I} = \dfrac{3.0}{0.45} = 6.67$ Ω and as $R_T = R + r$
	$6.67 = 5.0 + r$
Solve	$r = 1.67 = 1.7$ Ω

b The terminal potential of the battery in a closed circuit will be the emf of 3.0 V less the voltage drop across the internal resistor, r.

Given	$\varepsilon = 3.0$ V, $I = 0.45$ A, $R_L = 5.0$ Ω, $r = 1.7$ Ω
Unknown	$V = ?$
Equations	$V = \varepsilon - v$ and $V = IR$
Substitute	$V = 3.0 - (0.45 \times 1.767)$
Solve	$V = 2.249 = 2.2$ V

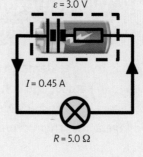

$\varepsilon = 3.0$ V

$I = 0.45$ A

$R = 5.0$ Ω

Investigating emf and internal resistance experimentally

The equation $V_{terminal} = \varepsilon - Ir$ is in the same form as the equation of a straight line: $y = c + xm$.

By decreasing the resistance of the external load, the current in the circuit increases. The terminal pd can be measured and a graph of $V_{terminal}$ against I plotted. A sample graph for the lamp circuit above (see page 257) is shown.

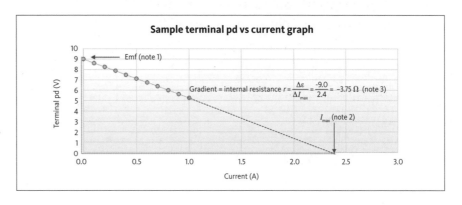

Sample terminal pd vs current graph

Emf (note 1)

Gradient = internal resistance $r = \dfrac{\Delta \varepsilon}{\Delta I_{max}} = \dfrac{-9.0}{2.4} = -3.75$ Ω (note 3)

I_{max} (note 2)

Terminal pd (V)

Current (A)

Key notes from the graph

1 When $I = 0$ A, the battery is in open circuit and from the equation above:

$$V_{terminal} = \varepsilon - 0r$$

The y-intercept represents the emf of the battery.

2 The x-intercept represents the **maximum current** of the battery. This will occur when the battery is being **short circuited** by connecting its terminals together with a conductor. As there is no resistance in the external load, it means that the voltage across the external load will be zero. From the equations above:

$$0 = \varepsilon - I_{max} r$$

Rearranging gives:

$$I_{max} = \frac{\varepsilon}{r}$$

All the battery's emf is being changed to thermal energy by the internal resistance and will result in the battery becoming very hot (this can be dangerous and should not be attempted experimentally).

3 Rearranging the equation above, $r = \dfrac{\varepsilon}{I_{max}}$, which means that the internal resistance can be found from the gradient of the graph, $m = \dfrac{\Delta \varepsilon}{\Delta I}$.

Worked example: Internal resistance and emf graphical analysis

Sam is trying to demonstrate that when resistors are added in parallel, the voltage remains the same across each branch, but as he adds identical resistors in parallel, the voltage across them drops as shown by the graph below.

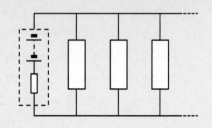

Sam realises that the power supply must have internal resistance that is affecting the output voltage.

a Using the graph, determine the emf of the power supply.

b Calculate the internal resistance of the power supply.

c Explain why the voltage across the external resistor network decreases as Sam adds more resistors in parallel.

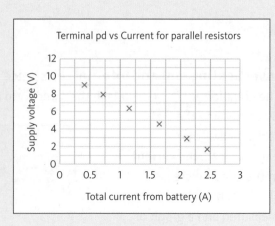

Terminal pd vs Current for parallel resistors

Solution

a By drawing a line of best fit through the points and extrapolating to the y-intercept, the emf of the supply is shown to be 10.5 V.

b The gradient of the graph gives the internal resistance: $-r = \dfrac{\Delta V}{\Delta I} = \dfrac{2.0 - 10.5}{2.4 - 0} = -3.54\ \Omega$

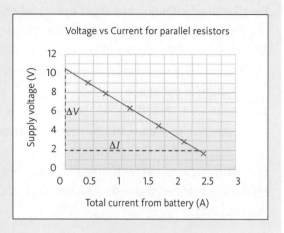

c As Sam adds each resistor in parallel with the first, he will be reducing the total resistance in the circuit. This causes the total current to increase and consequently there will be a greater voltage drop across the internal resistance, as $v = Ir$. The terminal pd $V = \varepsilon - Ir$, so as the voltage drop across the internal resistance increases, the terminal pd will decrease.

Internal resistance and the principle of conservation of energy

The energy changed in a source will be converted into different forms of energy inside the internal resistance and in the external load. The rate at which this occurs can be found by considering the power.

And as $P = I^2R$, this gives:

$$P_{supplied} = P_{dissipated\ in\ the\ external\ load} + P_{dissipated\ in\ the\ internal\ resistance}$$

The Principle of Conservation of Energy

Extension concepts for scholarship — Impedance matching

If a graph of power output against external resistance is plotted, then it reveals that the power output from the external load reaches a maximum then decreases again. The graph below demonstrates this for the sample circuit earlier (see page 257).

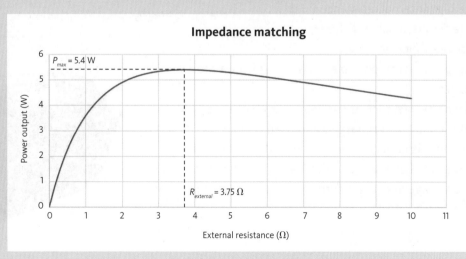

Key notes from the graph

1   The maximum power output occurs when **R = r**. This means that the maximum power output from the external load is **exactly half** the power supplied by the emf of the source.

2   When $R < r$, then most of the power is dissipated as thermal energy in the internal resistor. The pd across the external load will be low but there will be a large current due to the total resistance being low.

3   When $R > r$, then there will be a greater pd across the external load but the current will be lower due to the greater overall resistance.

Electrical engineers refer to this as impedance matching (impedance is another term for resistance that is discussed in AC circuits), and design electrical equipment to take account of this property.

Power sources and internal resistance

Power supplies

Basic power supplies can have a significant internal resistance that will affect the pd across the external load. More expensive designs use special feedback circuits to monitor their output and change accordingly to ensure a constant output.

Batteries

Primary batteries, such as alkaline batteries, can be **used once**, then must be thrown away as the chemical reactions are irreversible. During their operation the internal resistance increases, so the pd across the external load gradually decreases until the terminal pd is nearly zero (in closed circuit) and the battery is described as 'flat'. However, if the 'flat' battery is tested in open circuit, the terminal pd is the same as it was when the battery was new.

Due to the high internal resistance, primary batteries are unsuitable for high-drain devices such as mobile phones, laptops, cameras, portable tools, car starter motors and electric vehicles.

Secondary batteries, such as car (lead-acid) batteries, NiMH or NiCd batteries, can be **recharged** and used repeatedly. Their internal resistance is largely independent of the state of charge, but increases as the battery ages. Such batteries are able to deliver a more stable voltage and larger currents, so are suitable for high-drain applications.

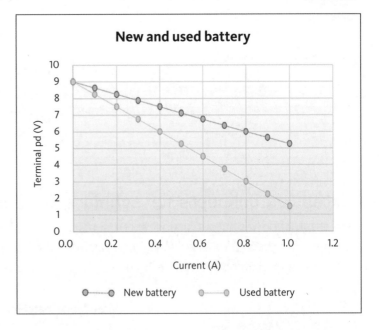

New and used battery

Solar cells

The internal resistance of a solar cell is not constant and increases as the current in the circuit increases.

The maximum power output is achieved just before the maximum current is reached, as there is still a significant terminal pd.

Light intensity and temperature both affect the power output of a solar cell.

- Reducing the light intensity significantly reduces the maximum terminal pd in open circuit.
- Increasing the temperature significantly reduces the maximum current in short circuit.

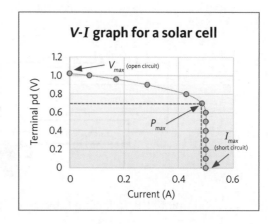

V-I graph for a solar cell

Exercise 6B

1 Describe the similarities and differences between the emf of a battery and the voltage drop across a resistor.

_____

_____

_____

_____

_____

_____

2 Describe the energy changes in the following devices and hence identify them as voltage drops or emfs.

Device	Energy change	V drop or emf
Generators	kinetic (mechanical) energy into electrical energy	
Loudspeakers	electrical energy into sound (mechanical) energy	
Lamps	electrical energy into heat and radiant energy	
Solar cells	radiant energy into electrical energy	
Microphones	sound (mechanical) energy into electrical energy	

3 A battery that is being used to light a torch rated 12 V and 5W is removed and used to operate a small handheld vacuum cleaner rated as 12 V and 90 W. The two circuits are shown below.

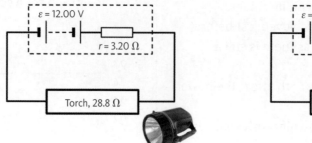

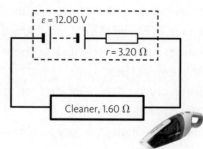

a Complete the following table by calculating all the values for the torch and vacuum cleaner.

Quantity	Torch	Cleaner
Total resistance in the circuit		
Current flowing around the circuit		
Terminal pd across the component		
Voltage across the internal resistance		
Power dissipated by the:		
emf of the battery		
internal resistor		
torch/cleaner		

b What will happen to the voltage drop across the internal resistance of a power supply if the resistance of the external load decreases?

_____

_____

_____

_____

_____

c Discuss how the battery will operate inside each device. In your answer you should consider the electrical quantities calculated in part **a**. (E)

_____

_____

_____

_____

_____

_____

_____

_____

_____

_____

_____

 ISBN: 9780170368179

Scholarship question

4   For the circuit to the right, prove that the power dissipated in the external load will be:

$$P = \frac{\varepsilon^2 R}{(R + r)^2}$$

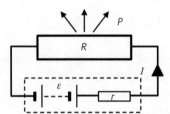

6.2 Kirchhoff's laws

The behaviour of the current and voltage in all series and parallel circuits can be understood by applying two basic laws — Kirchhoff's first (current) law and Kirchhoff's second (voltage) law. These laws are built on two fundamental conservation principles.

Kirchhoff's first (current) law (K1)

The Principle of **Conservation of Charge** states that:

> Electric charge cannot be created or destroyed, and that the net quantity of charge always remains the same.

When dealing with steady currents, this means that:

> The total current flowing into a junction equals the total current flowing out of that junction.

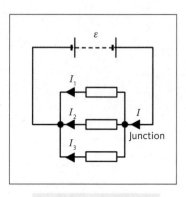

If we consider:

- current flowing **into** a junction as being **positive**, and
- current flowing **out** as **negative**,

$$I - I_1 - I_2 - I_3 = 0$$

then Kirchhoff's first law can be stated as:

$$\Sigma I = 0 \text{ A at a junction}$$

Kirchhoff's second (voltage) law (K2)

The Principle of **Conservation of Energy** states that:

> Energy cannot be created or destroyed but can be changed.

Around **any closed loop** in a circuit this means that:

> The total electrical potential energy supplied to the circuit must be equal to the total energy dissipated in the components as thermal, radiant or mechanical energy.

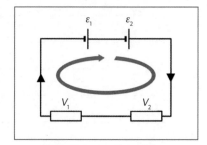

$$\varepsilon_1 + \varepsilon_2 - V_1 - V_2 = 0$$

If we consider:
- an increase in the voltage as being **positive**, e.g. battery emf
- a decrease in the voltage **as negative**, e.g. voltage drop across a resistor,

then Kirchhoff's second law can be stated as:

$$\Sigma \text{pd} = 0 \text{ around any closed loop}$$

Loop rules

The loops are a mathematical tool and don't exist so when drawing a closed loop, it doesn't matter in which direction the loop is drawn or the direction of any unknown currents as long as Kirchoff's laws and the V and I directions are applied consistently.

If a loop points in the opposite direction to the current flow then a component has the opposite effect, e.g. if a loop 'goes through' a resistor the 'wrong way' (opposite to the current) then there is an increase in the voltage.

Worked example: Kirchhoff's laws and series circuits

Two cells are connected in series with three resistors, as shown in the circuit opposite. Determine the size and direction of the current flowing around the circuit.

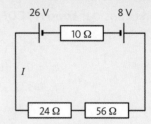

Solution

As the 26 V cell represents the largest emf in the loop causing the current to flow anticlockwise, take anticlockwise as the positive direction. The second cell must be negative, as it is opposing the current flow.

Draw a loop on the diagram showing the positive direction. As the loop passes each component, there is a voltage drop until the loop returns to 0 V at the negative terminal of the 26 V cell.

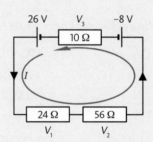

Given	$\varepsilon_1 = 26$ V (*anticlockwise*)
	$\varepsilon_2 = -8$ V (*anticlockwise*)
	$R_1 = 24\ \Omega$, $R_2 = 56\ \Omega$, $R_3 = 10\ \Omega$
Unknown	$I = ?$
Equations	$V = IR$ and $\Sigma pd = 0$ round any closed loop
	$\varepsilon_1 + \varepsilon_2 - V_1 - V_2 - V_3 = 0$ but as $V = IR$
	$\varepsilon_1 + \varepsilon_2 - IR_1 - IR_2 - IR_3 = 0$
Substitute	$26 + (-8) - (I \times 24) - (I \times 56) - (I \times 10) = 0$
	$18 - 90 \times I = 0$
Solve	$I = \dfrac{18}{90} = 0.20$ A (2 sf)

Kirchhoff's law applies to parallel circuits where there may be several closed loops which need to be considered.

Worked example: Kirchhoff's laws and parallel circuits

Two cells are connected in series with a resistor and then two resistors in parallel, as shown in the circuit opposite. Determine the size of the currents flowing around the circuit.

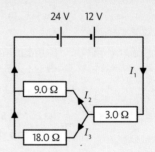

Solution

Both cells are causing the current to flow clockwise, so take clockwise as the positive direction. Using **Kirchhoff's first law** can relate the currents at the junction.

Given	
Unknown	$I_1 = ?$, $I_2 = ?$, $I_3 = ?$
Equations	$\Sigma I = 0$ A at a junction
Substitute	So at the junction $I_1 - I_2 - I_3 = 0$
Solve	$I_1 = I_2 + I_3$ or $I_2 = I_1 - I_3$ or $I_3 = I_1 - I_2$

There are three closed loops in this circuit. To make them easier to identify, label the corners and junction.

First consider **Loop ABCDFA**

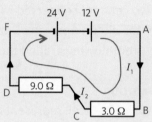

Given	ε_1 = 24 V (clockwise)
	ε_2 = 12 V (clockwise)
	$R_1 = 3\ \Omega,\ R_2 = 9\ \Omega$
Unknown	$I_1 = ?,\ I_2 = ?$
Equations	$V = IR$ and Σpd = 0 around any closed loop.
	$\varepsilon_1 + \varepsilon_2 - V_1 - V_2 = 0$ but as $V = IR$
	$\varepsilon_1 + \varepsilon_2 - I_1 R_1 - I_2 R_2 = 0$
Substitute	$24 + 12 - (I_1 \times 3) - (I_2 \times 9) = 0$ Rearranging:
	$36 = 3I_1 + 9I_2$ Dividing through by 3 gives:
Solve	$12 = I_1 + 3I_2$

Now consider **Loop ABCEFA**

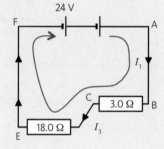

Given	ε_1 = 24 V (clockwise)
	ε_2 = 12 V (clockwise)
	$R_1 = 3\ \Omega,\ R_3 = 18\ \Omega$
Unknown	$I_1 = ?,\ I_3 = ?$
Equations	$V = IR$ and Σpd = 0 around any closed loop.
	$\varepsilon_1 + \varepsilon_2 - V_1 - V_3 = 0$ but as $V = IR$
	$\varepsilon_1 + \varepsilon_2 - I_1 R_1 - I_3 R_3 = 0$
Substitute	$24 + 12 - (I_1 \times 3) - (I_3 \times 18) = 0$ Rearranging:
	$36 = 3I_1 + 18I_3$ Dividing through by 3 gives:
Solve	$12 = I_1 + 6I_3$

Next combine Kirchhoff's first law, $I_2 = I_1 - I_3$, and the solution for Loop ABCDFA, and then use the solution with Loop ABCEFA to solve simultaneously for I_1. I_3 and I_2 can now be found by substituting back into the solution to Loop ABCEFA and Kirchhoff's first law, respectively.

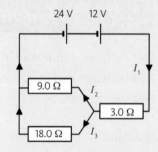

Given	Loop ABCDFA: $12 = I_1 + 3I_2$
	Loop ABCEFA: $12 = I_1 + 6I_3$
	$K_1: I_2 = I_1 - I_3$
Unknown	$I_1 = ?$
Equations	Combining $12 = I_1 + 3I_2$ and $I_2 = I_1 - I_3$ gives
	$12 = I_1 + 3(I_1 - I_3)$
	$12 = 4I_1 - 3I_3$ for Loop ABCDFA
Substitute	To solve, simultaneously multiply Loop ABCDFA by 2 and then add the two solutions together.
	Loop ABCDFA: $24 = 8I_1 - 6I_3$
	Loop ABCEFA: $12 = I_1 + 6I_3$
	$36 = 9I_1$

Solve

$$I_1 = \frac{36}{9} = \textbf{4.00 A}$$

From Loop ABCEFA: $12 = I_1 + 6I_3$

so $I_3 = \frac{12 - 4}{6} = \textbf{1.33 A}$

From K_1: $I_2 = I_1 - I_3$ so $I_2 = 4.00 - 1.33 = \textbf{2.67 A}$

The problem above was solved without considering the third closed loop CEDC. The voltage at C can be found by considering the total emf and the voltage drop across the 3.0 Ω resistor and is given by:

$$V_C = \varepsilon_1 + \varepsilon_2 - I_1 R_1$$
$$V_C = 36 - 4.00 \times 3$$
$$V_C = 24 \text{ V}$$

The voltage at E and D is zero, as they are both in contact with the negative terminal of the 24 V cell. Following the loop from C to E through the 18 Ω resistor results in a drop of 24 V, but now as the loop goes from D to C it is moving in the **opposite direction to the current** through the 9.0 Ω resistor and the voltage increase from zero to 24 V, consequently the **voltage drop** across the 9.0 Ω **is negative** because we observed a voltage gain. Hence Kirchhoff's second law $\Sigma\varepsilon = \Sigma V$ for Loop CEDC looks like this:

$$0 = I_3 R_3 + I_2 R_2$$
$$0 = 1.33 \times 18 + (-2.67) \times 9.0$$
$$0 = 24 - 24 \qquad \text{which agrees with Kirchhoff's second law.}$$

Exercise 6C

1 State Kirchhoff's first law. (A)

_____

_____

_____

_____

2 State Kirchhoff's second law. (A)

_____

_____

_____

_____

3 Charge and current are not the same thing — so explain using words and mathematical expressions how Kirchhoff's first law is an expression of the principle of conservation of charge. (A)

_____

_____

_____

_____

_____

4 Voltage and energy are not the same thing — so explain using words and mathematical expressions how Kirchhoff's second law is an expression of the principle of conservation of energy. (A)

_____

_____

_____

_____

_____

5 Use knowledge of Ohm's and Kirchhoff's laws to determine the missing values in the following circuits.

a (A)

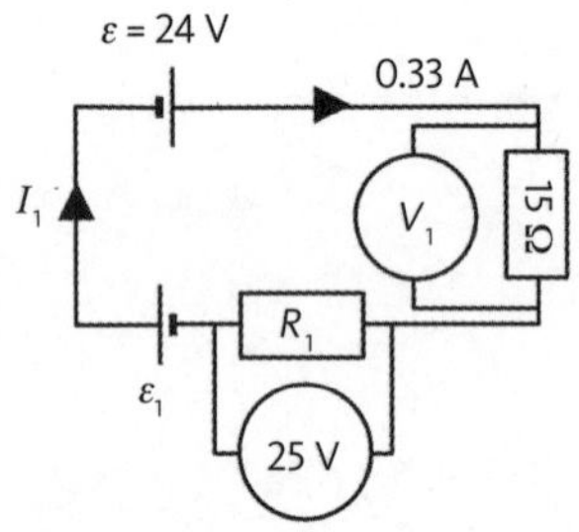

b (A)

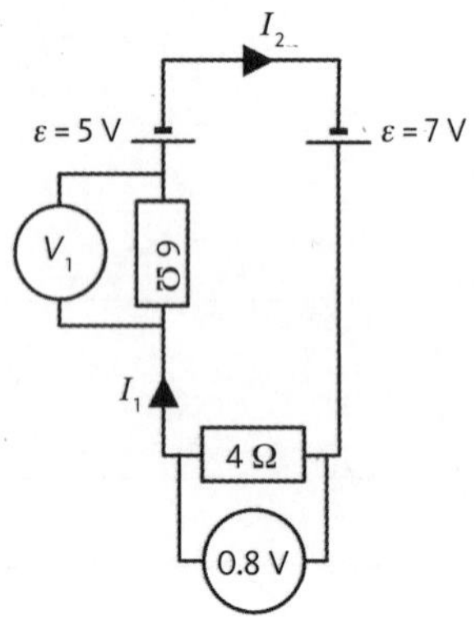

c   (M)

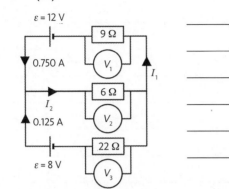

$\varepsilon = 12$ V

9 Ω

0.750 A V_1

I_1

6 Ω

I_2

0.125 A V_2

22 Ω

$\varepsilon = 8$ V

V_3

6   Use Ohm's and Kirchhoff's laws to determine the missing values in the following circuit. You may find it helpful to follow the steps below.

- Start by labelling the corners and junctions: A, B, C, …
- Draw in current directions if they are not given.
- Identifying the closed loop(s).
- Using the loop rules, annotate + and – signs to the cells and resistors.
- Apply Kirchhoff's first and second laws to solve.

a   Find I_1, I_2, I_3, and the voltage across each resistor. (E)

$\varepsilon_1 = 18$ V

9.0 Ω

I_1

I_3

3.0 Ω

I_2

1.5 Ω

$\varepsilon_2 = 6.0$ V

b Find I_3, R_2, R_3, and the voltage across each resistor. (E)

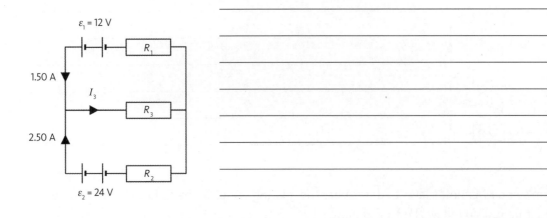

c Find I_1, I_2, I_3, and the voltage across each resistor. (E)

d For each example above, write the voltage drop across each resistor on the diagrams.

    i What can you say about the emfs and voltage losses around each closed loop? (A)

    ii What can you say about the voltage across each branch? (A)

7 Michelle likes to play games on her iPhone 4S. The battery is rated as 3.70 V and 5.3 W hr.

   a Identify what quantity is measured in 'W hr' and convert it to an alternative unit. (A)

_____

_____

_____

Her iPhone requires 0.88 W of power to operate when playing games.

   b How long can Michelle play games until her battery 'runs out'? (A)

_____

_____

_____

   c Show that the current drawn by the phone while Michelle plays games is 0.24 A. (A)

_____

_____

_____

The internal resistance of the lithium ion battery is 0.40 Ω and doesn't change while the phone is in operation.

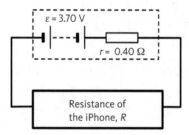

   d Calculate the total resistance of the iPhone components when Michelle is playing games on her phone. (A)

_____

_____

When the phone finally runs out of power, Michelle connects her iPad and iPhone to a double charging unit as shown in the circuit diagram at right.

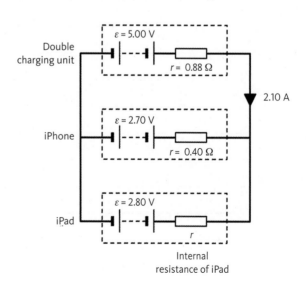

e Calculate the internal resistance of the iPad. (M)

_____

_____

_____

_____

_____

Scholarship questions

8 Portable electric devices such as digital cameras have one set of batteries but the different components inside require a range of voltages. Potential dividers are used to provide the appropriate voltage to each component. The circuit diagram below shows an example of a potential divider circuit.

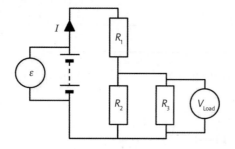 

a Prove that the output voltage across the load is given by the relationship

$$V_{Load} = \frac{\varepsilon R_2 R_3}{R_1 R_2 + R_2 R_3 + R_1 R_3}$$

_____

_____

_____

_____

_____

_____

_____

_____

_____

_____

_____

_____

_____

_____

_____

_____

_____

 ISBN: 9780170368179

_____

_____

_____

_____

_____

_____

_____

_____

_____

_____

To determine the emf of a battery, an analogue voltmeter can be placed across it while in open circuit as shown below. Real analogue voltmeters do not have an infinite resistance and allow a small current to flow through them. The voltmeter can be modelled by an ideal voltmeter with an internal resistance of 1.000 kΩ.

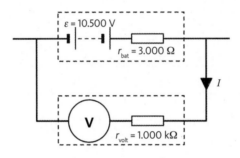

b   Determine the emf of the battery when using the real voltmeter. Explain whether the error is significant.

_____

_____

_____

_____

_____

_____

_____

_____

_____

_____

Potential dividers circuits can also be used in battery-testing devices to measure the emf of a battery. The circuit is shown below with a variable resistor to adjust the voltage across the cell being tested and a galvanometer (G) being used to detect very small currents.

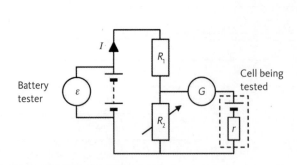

The circuit is described as 'balanced' when the galvanometer reads zero and the emf can be determined.

c   Explain what is meant by the term 'balanced' in this context and how the circuit enables the emf of the battery to be determined.

_____

_____

_____

_____

_____

_____

_____

_____

_____

_____

_____

_____

_____

_____

6.3 Capacitors in DC circuits

Capacitors

Capacitors are designed to store electrical energy. They are basically made from two conductors, such as a pair of parallel metal plates, separated by an insulator.

The insulator is called a dielectric and can be made from any non-conducting material, such as waxed paper.

Due to the insulating properties of the dielectric, current cannot flow through a capacitor, but charges can flow onto one plate and off the other plate. As the amount of charge on each plate increases, the voltage across the capacitor increases.

When the voltage (V) across the capacitor in the example opposite reaches the same value as the supply voltage (ε), no more charge is able to flow on to the capacitor and the charging current stops.

Capacitors store energy in the electric field between the plates by keeping electrical charge on their plates so can also be thought of as storing charge, however the net charge on a capacitor is always zero as the amount of positive charge on one plate of the capacitor is equal to the negative charge on the other plate.

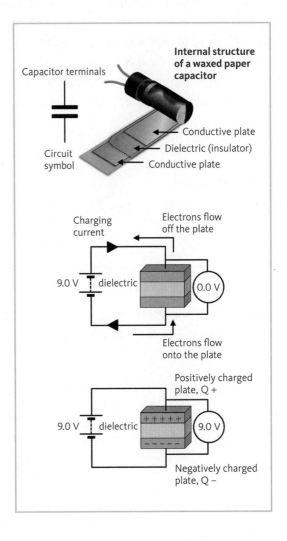

Capacitance (C)

Capacitance (C) is a measure of a capacitor's ability to store charge and is measured in farads (F). The charge on the plate of a capacitor is directly proportional to the potential difference across it, so the gradient of a graph of charge against potential difference gives the capacitance.

$$\text{capacitance } (F) = \frac{\text{charge } (C)}{\text{voltage } (V)}$$

Expressed mathematically: $C = \dfrac{Q}{V}$ or more commonly as: $Q = CV$

where 1 F = 1 CV⁻¹. In practice, most capacitors used in electronics are very small and so are measured in microfarads (μF, x 10^{-6}), nanofarads (nF, x 10^{-9}) or picofarads (pF, x 10^{-12}).

Energy stored in a capacitor, E

As soon as a capacitor starts charging, work must be done to push more electrons onto the negative plate against the repulsive force caused by the electrons already there. On the positive plate, work must be done to remove electrons against the force of attraction.

The work that is done is stored as electrical potential energy in the capacitor and can be determined by finding the area under a graph of charge against potential difference.

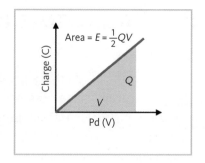

$$E = \frac{1}{2} QV$$

which states that:

$$\text{energy (J)} = \frac{1}{2} \times \text{charge (C)} \times \text{potential difference (V)}$$

If a total charge (Q) moves around the circuit during this charging process, then the battery will do work (ΔE), where: $\Delta E = Q\varepsilon$.

But the capacitor will store $\Delta E = \frac{1}{2} Q\varepsilon$, which is half the energy changed by the battery. This means that half the energy must be dissipated by the resistance in the circuit.

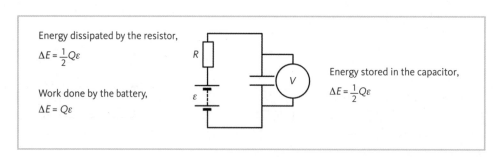

By combining the capacitance formula $Q = CV$ and the energy formula $E = \frac{1}{2} QV$, the energy can be expressed in terms of the capacitance as:

$$E = \frac{1}{2}CV^2$$

which reveals the proportional relationship between energy and capacitance.

 Worked example: Flashy photography skills

A camera flash uses a capacitor with a paper dielectric to store sufficient charge to 'fire' the flash. The 160 µF capacitor discharges in 1.74 ms causing a flash of light. The flash unit has a power rating of 5.00 kW.

a Show that the energy stored in the capacitor is 8.7 J.

b Calculate the voltage across the plates of the capacitor.

Solution

a

Given	$P = 5.00 \times 10^3$ W
	$t = 1.74 \times 10^{-3}$ s
Unknown	$E = ?$
Equations	$E = Pt$
Substitute	$E = 5.00 \times 10^3 \times 1.74 \times 10^{-3}$
Solve	$E = 8.7$ J (2 sf)

b

Given	$E = 8.7$ J
	$C = 1.60 \times 10^{-6}$ F
Unknown	$V = ?$
Equations	$Q = CV$ and $E = \dfrac{1}{2}QV$ combined give $E = \dfrac{1}{2}CV^2$
Substitute	$8.7 = \dfrac{1}{2} \times 160 \times 10^{-6} \times V^2$
	$\dfrac{8.7 \times 2}{160 \times 10^{-6}} = V^2 = 108{,}750$
	$V = \sqrt{108\,750}$
Solve	$V = 329.8$ V $= 330$ V (2 sf)

Exercise 6D

1 A 680×10^{-12} F capacitor is connected to a 5.0 V power supply and fully charged. Calculate the charge on the capacitor.

_____

_____

2 When a capacitor is connected to a 12.0 V supply, it is able to store 264×10^{-6} C on its plates.

    a Calculate the size of its capacitance. (A)

_____

_____

    b Calculate the energy stored by the capacitor. (A)

_____

_____

3 A fully charged capacitor stores 180.6 J of energy when connected to a 630 V power supply.

    a Calculate the charge stored on the plates of the capacitor. (A)

_____

_____

b Calculate the capacitance of the capacitor. (M)

_____

_____

4 A fully charged 528 nF capacitor can store ±264 μC on its plates when connected to a power supply.

   a Calculate the voltage across the plates of the capacitor. (A)

_____

_____

   b State the net charge on the capacitor and explain your answer. (M)

_____

_____

_____

5 A 370 μF capacitor is able to store 3.33 mC of charge when attached to a battery.

   a Show that the emf of the battery when the capacitor is fully charged is 9.0 V. (A)

_____

_____

   b Calculate the energy stored on the capacitor and compare it to the energy given out by the battery. (M)

_____

_____

_____

_____

_____

_____

_____

   c Explain what will happen to the charge and energy stored in the capacitor if the battery voltage is tripled (3x greater)? (M)

_____

_____

_____

_____

_____

6 A defibrillator, which is used to return a patient's heart to a normal rhythm, uses a 32.0 μF capacitor charged to 800.0 V. When discharged, it delivers a short burst of current for 3.0 ms.

   a Show that 25.6 mC of charge are stored on the capacitor. (A)

_____

_____

b By first considering the average current that flows through the patient's body in 3.0 ms, calculate the resistance of the patient's body to the flow of charge. (M)

_____

_____

_____

_____

c Explain why it is important to clean the patient's chest and apply a conducting gel before attaching the defibrillator pads. (E)

_____

_____

_____

_____

_____

_____

_____

7 A wind turbine uses an oil filled capacitor with a rating of 400 uF $\pm$ 5% 1100 V DC. The '$\pm$ 5%' refers to the accuracy of the measurement of the capacitance.

a Calculate the maximum and minimum values of the capacitance.

_____

_____

To determine the actual capacitance, the capacitor is tested and a graph of the results plotted opposite.

b Using the graph, determine the actual capacitance of the capacitor. (A)

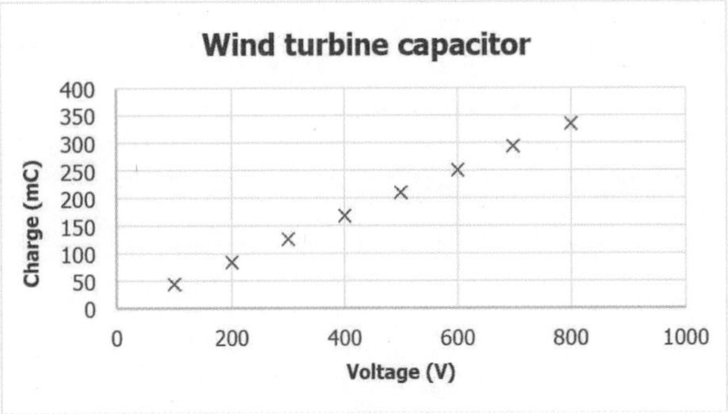

_____

_____

_____

Wind turbine capacitor

c Using the graph or otherwise, show that 75 J of energy is stored on the capacitor when the turbine creates a potential difference of 600 V across its plates. (A)

_____

_____

d Calculate the average power output from the capacitor if it was accidentally discharged by a short circuit in a time of 5.0 ms.(A)

_____

_____

The capacitor is damaged by the short circuit and is replaced by a 200 μF capacitor.

e On the graph above, sketch a line to show how the charge and voltage will change and discuss the impact on the total energy stored. (E)

_____

_____

_____

_____

_____

_____

_____

_____

_____

Scholarship question

8 By considering fundamental quantities and SI units, prove that $1\ F = 1\ A^2\ s^4\ m^{-2}\ kg^{-1}$.

_____

_____

_____

_____

_____

_____

_____

_____

_____

_____

_____

_____

Parallel plate capacitor properties

The simplest capacitor is made up of two parallel conducting plates of area A separated by a distance d. The capacitance is given by the equation:

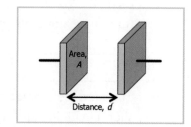

$$C = \frac{\varepsilon_0 A}{d}$$

where $\varepsilon_0 = 8.854 \times 10^{-12}$ F m^{-1} is the permittivity of free space and is a fundamental constant which describes how the presence of a vacuum affects the strength of the electric field between the plates. (Note: the permittivity of air is considered to be the same as that of a vacuum.)

 ISBN: 9780170368179

Dielectric materials
The capacitance can be increased by adding a non-conducting dielectric material.
- The electrons in an insulating dielectric are tightly bound to the nuclei of the molecule, and are distributed symmetrically around them.
- An electric field between the plates disturbs the symmetrical nature of the electron's motion and the molecule becomes polarised, with one end becoming positive and the other end becoming negative.
- As a consequence, an electric field forms inside the dielectric and acts in the opposite direction to the field due to the external plates.
- The two fields combine and the net electric field strength between the plates decreases.

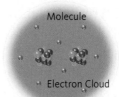

Molecule

Electron Cloud

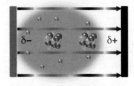

Dielectric

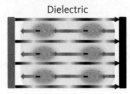

As $V = Ed$, a decrease in the net electric field results in a similar decrease in the voltage between the plates. If the capacitor is isolated so that that charge on the capacitor is fixed, then the introduction of a dielectric between the plates will cause the capacitance to increase as $C = \dfrac{Q}{V}$.

Relative permittivity (ε_r)
The relative permittivity (ε_r) is the factor by which the capacitance is increased by the presence of a dielectric. The relative dielectric constant, ε_r is defined as:

$$\varepsilon_r = \frac{\text{capacitance of parallel plate capacitor with dielectric (F)}}{\text{capacitance of an identical capacitor with vacuum between plates (F)}}$$

Since the dielectric increases the capacitance of the capacitor, the value of ε_r is always greater than 1.0. Some examples are given in the table at right. Relative permittivity has no unit.

Dielectric material	ε_r
Air	1.000
Polythene	2.25
Paper	3.85
Glass	4–10
Water*	80
Barium titanate*	1250–10 000

*Note: Polar molecules
Some molecules, such as water, are polarised even without the action of an external electric field. If a material which contains polar molecules is used as the dielectric, then the molecules become aligned in the direction of the electric field resulting in a very large relative permittivity and a significant increase in the capacitance of the capacitor.

With a dielectric inserted between the plates, the parallel plate capacitor equation becomes:

$$C = \frac{\varepsilon_0 \varepsilon_r A}{d}$$

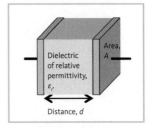

which states:

$$\text{capacitance (F)} = \frac{\text{permittivity of free space (F m}^{-1}) \times \text{relative permittivity} \times \text{area (m}^2)}{\text{separation of the plates (m)}}$$

Worked example: Waxing lyrical

Eddie is testing the effect of dielectrics on capacitance. He places two aluminium plates 30.0 cm long and 30.0 cm wide on an insulated stand at a distance of 1.00 mm and applies a voltage of 18 V across the plates. Eddie measures the capacitance then slides a piece of waxed paper into the space between the plates **while it is still connected** to the supply and observes the capacitance increase to 3.07 nF. Permittivity of free space, $\varepsilon_0 = 8.854 \times 10^{-12}$ F m^{-1}.

a Calculate the permittivity of the waxed paper dielectric. (M)

b Calculate the charge stored on the capacitor. (A)

c Explain why inserting the waxed paper affected the capacitance of the capacitor. (E)

Solution

a Watch out for unit conversions. The question contains cm, mm and nF.

Given	$l = 0.300$ m, $w = 0.300$ m, $C = 3.07 \times 10^{-9}$ F, $\varepsilon_0 = 8.854 \times 10^{-12}$ F m^{-1}, $d = 0.00100$ m
Unknown	$\varepsilon_r = ?$
Equations	$C = \dfrac{\varepsilon_0 \varepsilon_r A}{d}$ and $A = l \times w$
Substitute	$3.07 \times 10^9 = \dfrac{8.854 \times 10^{-12} \times \varepsilon_r \times 0.30 \times 0.30}{0.00100}$ $\dfrac{3.07 \times 10^{-9} \times 0.00100}{8.854 \times 10^{-12} \times 0.30 \times 0.30} = \varepsilon_r$
Solve	$\varepsilon_r = 3.85$ (3 sf)

b

Given	$V = 18$ V, $C = 3.07 \times 10^{-9}$ F
Unknown	$Q = ?$
Equations	$Q = CV$
Substitute	$Q = 3.07 \times 10^{-9} \times 18$
Solve	$Q = 55.26 \times 10^{-9} = 55$ nC (2 sf)

c Inserting the dielectric between the plates causes the electric field to decrease and hence the voltage across the plates. However, as the capacitor is still connected to the battery a current will flow increasing the charge and returning the voltage across the plates to its original value. Increasing the number of charges on each capacitor plate results in an increase in the capacitance, as $C = Q/V$.

ISBN: 9780170368179

Exercise 6E

In the following questions, take the permittivity of free space as $\varepsilon_0 = 8.854 \times 10^{-12}$ F m^{-1}.

1 A capacitor is constructed using two aluminium plates, each with a surface area of 0.160 m^2. They are placed either side of a sheet of polystyrene, which is 5.00 mm thick and has a relative permittivity of $\varepsilon_r = 2.7$. Calculate the capacitance. (A)

_____

_____

_____

2 A sheet of Pyrex (a type of glass) has a relative permittivity of 4.7 and is placed between two metal plates, each with an area of 0.25 m^2, to make a capacitor with a capacitance of 5.20 nF. Determine the distance between the plates. (A)

_____

_____

_____

3 The plates in a 620 nF capacitor are separated by a 0.80 mm-thick layer of barium strontium titanate. Barium strontium titanate is an extremely effective dielectric and has a relative permittivity of 500.

    a Show that the surface area of the plates is 0.11 m^2. (A)

_____

_____

_____

    b If the plates are circular, calculate the radius of the plates. (A)

_____

_____

_____

4 Two metal foil sheets 1.00 cm wide by 12.00 cm long are separated by a thin dielectric only 0.10 mm thick to form a capacitor. The capacitance is 160 pF. Calculate the relative permittivity of the dielectric material. (M)

_____

_____

5 The capacitor in a camera flash stores 72.0 mC of charge at a voltage of 400 V. The plates on the capacitor have an area of 0.68 m^2 and the dielectric has relative permittivity of 1200.

    a Show that the plates are separated by a distance 40 μm. (M)

_____

_____

_____

_____

_____

b Calculate the electric field strength between the plates. (A)

_____

_____

6 The graph below shows how water's relative permittivity decreases as the temperature of the water increases. This property can be used to measure the temperature of water in a tank with a volume of 0.192 m³.

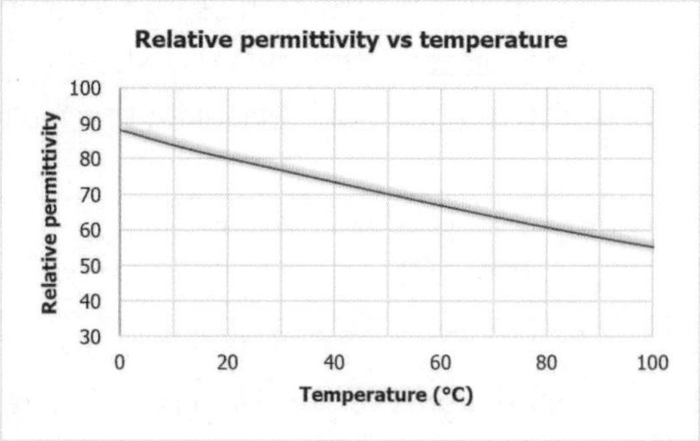

With the aid of the graph, determine the temperature of the water in the tank if the plates are 40 cm apart and the capacitance is 743.7 pF. (M)

Relative permittivity vs temperature

_____

_____

_____

7 A parallel plate capacitor is charged up, then isolated from the battery and the earth. The two metal plates are then moved in opposite directions until only a third of each plate overlaps with the other plate. Explain how this affects the capacitance and the charge on the capacitor. (M)

_____

_____

_____

_____

_____

8 Discuss the effect of doubling the distance between the plates of a parallel plate capacitor on the energy stored by the capacitor when it is isolated and when it is connected to a battery. (E)

_____

_____

_____

_____

_____

_____

_____

_____

_____

_____

_____

_____

_____

Scholarship questions

9   A sensitive electronic balance can be built using a capacitor as shown below.

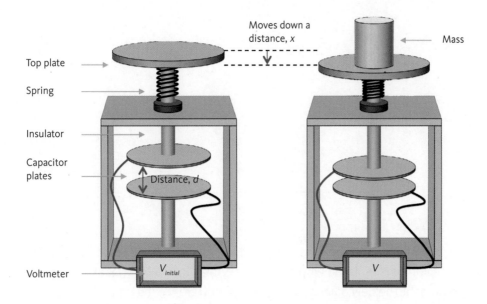

The capacitor is initially charged by connecting it to a power supply, then isolated from the supply and the earth. An ideal voltmeter is connected across the plates of the capacitor to measure the potential difference between the plates.

Placing a mass onto the top plate causes the spring (with a spring constant k) to be compressed and the upper capacitor plate to move downwards a distance x.

a   Prove that the potential difference between the plates of the capacitor when the mass is added is given by the formula

$$V = \frac{Qd}{\varepsilon_0 A} - \frac{Qmg}{\varepsilon_0 Ak}$$

b On the graph below, draw a line (label it (b)) to show how the potential difference on the plates varies with the mass placed on the top plate. The initial charging voltage, V_c, has been labelled on the graph to help you.

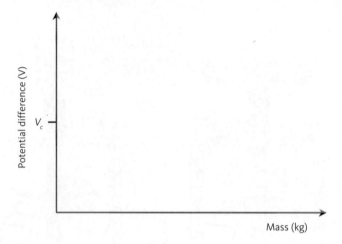

c Explain what will happen to the sensitivity and the measurement range of the electronic balance if the spring is replaced with a weaker spring. Draw a line (label it (c)) on the graph above to represent the new balance's response to difference masses.

_____

_____

_____

_____

_____

_____

_____

_____

d With reference to the formula, discuss an alternative change that could be made to improve the electronic balance and how it impacts on the sensitivity and range. Draw a line (label it (d)) on the graph above to represent the new balance's response to difference masses.

_____

_____

_____

_____

_____

_____

_____

_____

_____

_____

_____

_____

_____

_____

_____

_____

_____

_____

Capacitor combinations

When a single capacitor of the correct value is not available, a set of capacitors can be combined in series or parallel to create an equivalent capacitor of the desired value.

Capacitors connected in series

When capacitors are placed in series the charge, Q on each capacitor is the same regardless of the size of the capacitor. This is because the current is the same at every point in a series circuit and so the number of electrons flowing onto the negative plate of the first capacitor must flow off its positive plate and onto the next capacitor's negative plate, and so on to the last capacitor in the series.

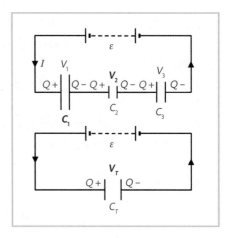

This means that the **total charge** must also be the **same as** the **charge on each capacitor**, as the outside plates on the first and last capacitors have a charge of $Q +$ and $Q -$.

The equivalent capacitance is given by the formula:

$$\frac{1}{C_T} = \frac{1}{C_1} + \frac{1}{C_2} + \frac{1}{C_3} \dots$$

Capacitors connected in parallel

When capacitors are placed in parallel, the voltage, V, across each capacitor is the same regardless of the size of the capacitor.

This means that the larger the capacitor, the greater the charge that can be stored for the same voltage, as $Q = CV$ and the total charge will be the sum of the charges on each capacitor.

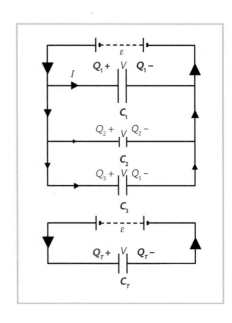

The equivalent capacitance is given by the formula:

$$C_T = C_1 + C_2 + C_3 \dots$$

Adding capacitors in parallel is like adding the areas of their plates to form one large plate.

Parallel and series circuits combined

Step 1: Combine any capacitors in series on the main circuit or in parallel branches.

Step 2: Combine the equivalent capacitors in parallel.

Step 3: Combine these equivalent capacitors in series.

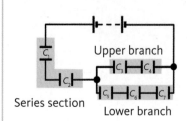

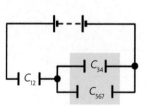

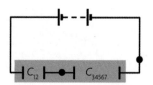

Worked example: Combining capacitors

Three capacitors of capacitance 16 μF, 18 μF and 30 μF are connected to a 9.0 V power supply as shown.
Calculate:

a The total capacitance of the circuit.

b The total charge.

c The charge on the 16 μF capacitor.

d The voltage across the 18 μF capacitor.

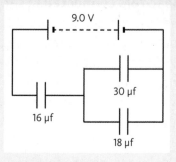

Solution

a

Given	$C_{series} = 16 \times 10^{-6}$ F, $C_{para\,1} = 30 \times 10^{-6}$ F, $C_{para\,2} = 18 \times 10^{-6}$ F
Unknown	$C_T = ?$
Equations	$C_{combined\,parallel} = C_{para\,1} + C_{para\,2}$
	$\dfrac{1}{C_T} = \dfrac{1}{C_{series}} + \dfrac{1}{C_{combined\,parallel}}$
Substitute	$C_{combined\,parallel} = 30 \times 10^{-6} + 18 \times 10^{-6}$
	$C_{combined\,parallel} = 48 \times 10^{-6}$ F
	$\dfrac{1}{C_T} = \dfrac{1}{16 \times 10^{-6}} + \dfrac{1}{48 \times 10^{-6}}$
Solve	$C_T = 12 \times 10^{-6}$ F = 12 μF (2 sf)

b

Given	$C_T = 12 \times 10^{-6}$ F, $\varepsilon = 9.0$ V
Unknown	$Q_T = ?$
Equations	$Q_T = C_T \varepsilon$
Substitute	$Q_T = 12 \times 10^{-6} \times 9$
Solve	$Q_T = 108 \times 10^{-6}$ C = 110 μC (2 sf)

c The charge on the 16 μF capacitor will be 110 μC, as the charge on capacitors connected in series is the same at every point and the same as the total charge. This means that ...

   ISBN: 9780170368179

d ... the charge shared between the combined 18 µF and 30 µF capacitors will also be 110 µC.

Given $C_{combined\ parallel}$ = 48 x 10⁻⁶ F (from **a**)

                  $Q_{combined\ parallel}$ = 108 x 10⁻⁶ C (from **b** and **c**)

Unknown $V_{combined\ parallel} = V_{18}$ = ? (parallel network)

Equations $V_{18} = \dfrac{Q_{combined\ parallel}}{C_{combined\ parallel}}$

Substitute $V_{18} = \dfrac{108 \times 10^{-6}\ C}{48 \times 10^{-6}\ F}$ = 2.25 V

Solve V_{18} = 2.3 V (2 sf)

Exercise 6F

1   Calculate the equivalent capacitance when connecting three capacitors with values 4 µF, 5µF and 20 µF:

    a   in parallel

    _____

    b   in series

    _____

    _____

2   Determine the possible equivalent capacitances that could be made using three 6 µF capacitors. Draw each circuit arrangement in the space provided.

3 An 8.0 µF capacitor and a 10.0 µF capacitor are connected in parallel with a 12 V supply.

   a Show that the total capacitance of the circuit is 18 µF. (A)

_____

   b Calculate the total charge. (A)

_____

   c Calculate the charge on the 8 µF capacitor when fully charged. (A)

_____

   d Calculate the charge on the 10 µF capacitor when fully charged. (A)

_____

   e Explain the relationship between the total charge and the charge on each individual capacitor. (M)

_____

_____

_____

_____

_____

4 Two capacitors are connected in series with an 18.0 V supply and left until the flow of charges stops.

   a Calculate the total capacitance when $C_1 = 60$ µF and $C_2 = 300$ µF. (A)

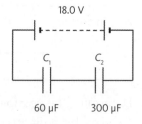

_____

_____

_____

   b Calculate the charge stored on each capacitor and explain your answer. (M)

_____

_____

_____

   c Calculate the total energy stored by the capacitor. (A)

_____

_____

_____

5 Three capacitors of capacitance 28 µF, 14 µF and 70 µF are connected to a 12.0 V power supply as shown.

   a Calculate the total capacitance of the circuit. (A)

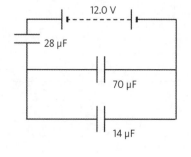

_____

_____

_____

_____

 ISBN: 9780170368179

b Show that the voltage across the parallel network is 3.0 V. (M)

_____

_____

_____

_____

c Calculate the charge on the 14 µF and 70 µF capacitors. (A)

_____

_____

d Calculate the energy stored on the 14 µF and 70 µF capacitors. (A)

_____

_____

_____

When the capacitors have finished charging, the parallel
network is removed from the circuit.

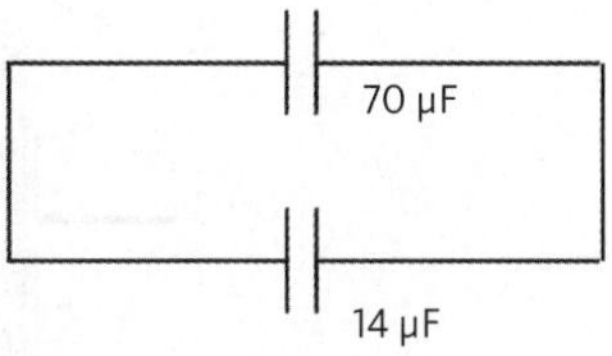

e Explain what will happen to the potential difference,
charge and energy of the capacitors when they are
removed from the circuit. (M)

70 µF

14 µF

_____

_____

_____

_____

6 Concrete takes several days to harden and cure. One method of monitoring the process is to
measure the relative permittivity of the concrete, which gradually decreases as the concrete
dries. This can be done by placing a sample of the concrete in a test capacitor. The test capacitor
has a capacitance of 55.33 pF when empty and 940.7 pF when filled with wet concrete.

a Show that the relative permittivity of wet concrete is 17. (A)

_____

_____

_____

The test capacitor has a volume 0.010 m³ with metal plates on either side separated by a
distance of 4.00 cm. The sample is left to dry out, and is found to have a new capacitance of
332.0 pF.

b Calculate the relative permittivity of dry concrete. (M)

_____

_____

_____

c Explain why the capacitance decreases as the concrete dries out and state any assumptions you have made. (M)

_____

_____

_____

_____

_____

_____

_____

The test capacitor is not cleaned out properly and the bottom half is still full of dry concrete when new wet concrete is poured on top. The test capacitor now behaves as though there are two capacitors in parallel as shown below.

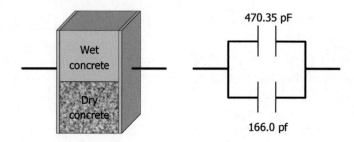

d Determine the effective relative permittivity of the mix and hence explain why it is important that the test capacitor is thoroughly clean before retesting occurs. (E)

_____

_____

_____

_____

_____

_____

_____

_____

_____

_____

_____

_____

_____

_____

Scholarship questions

7 Two capacitors are connected in parallel to give an equivalent capacitance of 9 F. When connected in series, they give an equivalent capacitance of 2 F. Determine the capacitance of the two capacitors.

_____

_____

_____

_____

_____

_____

_____

8 Once a capacitor in a series circuit is fully charged, the current stops flowing.

a Explain why the current stops flowing. (M)

_____

_____

_____

_____

_____

_____

_____

_____

_____

_____

A 560 µF capacitor (C_1) is fully charged using a 12 V supply, then connected in parallel with an uncharged 280 µF capacitor (C_2) as shown below.

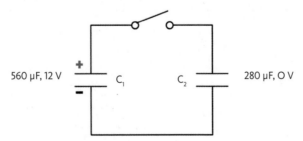

b Discuss, with the aid of calculations, what will happen when the switch is closed. In your answer you need to consider the following:
- The total charge in the circuit.
- The total capacitance of the circuit.
- The potential difference across each capacitor.
- The charge on each capacitor.
- The initial and final energies stored by the capacitors. (E)

ISBN: 9780170368179

Charging and discharging capacitors

The flexibility to alter the time to charge and discharge a capacitor is fundamental to their use in electronic circuits, such as timing circuits, flash guns, and power smoothing.

Charging

When a capacitor is connected to the supply, electrons start to flow onto one plate and off the other plate. This causes one plate to become negatively charged and the other plate to become positively charged.

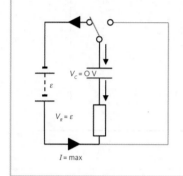

I = max

As the potential difference between the plates increases, the negative plate resists the flow of additional electrons onto it and as a consequence the rate at which charge flows on and off the capacitor rapidly decreases.

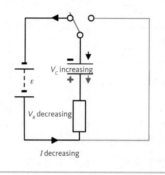

I decreasing

The current in the circuit continues to decrease until the potential difference across the plates of the capacitor is equal to the emf of the supply, at which time the charges stop moving and the current falls to zero.

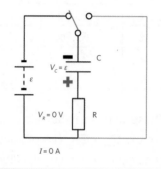

I = 0 A

The graphs below show how the current in the circuit, potential difference across the capacitor and resistor, and the charge on the capacitor change with time during the charging process.

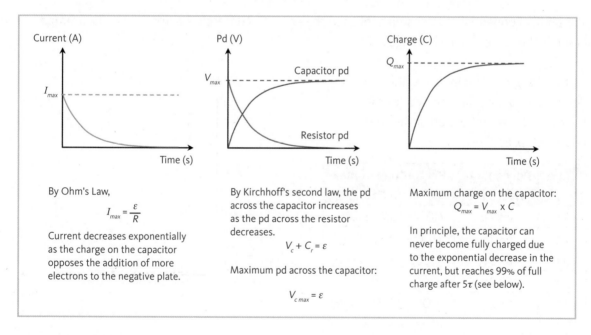

By Ohm's Law,

$$I_{max} = \frac{\varepsilon}{R}$$

Current decreases exponentially as the charge on the capacitor opposes the addition of more electrons to the negative plate.

By Kirchhoff's second law, the pd across the capacitor increases as the pd across the resistor decreases.

$$V_c + C_r = \varepsilon$$

Maximum pd across the capacitor:

$$V_{c\,max} = \varepsilon$$

Maximum charge on the capacitor:

$$Q_{max} = V_{max} \times C$$

In principle, the capacitor can never become fully charged due to the exponential decrease in the current, but reaches 99% of full charge after 5τ (see below).

Time constant, τ (tau)

The time it takes to charge (or discharge) a capacitor depends on:

- the size of the capacitor and
- the resistance in the circuit,

and is described using the **time constant**, τ, where:

time constant (s) = resistance (Ω) x capacitance (F)

Expressed mathematically:

$$\tau = RC$$

The time constant represents the time taken for the charge on a capacitor to increase to 63.2%, resulting in an exponential relationship that means that a capacitor can never be 100% charged. The table and graph show the relationship between charge and time constants.

Time	Percentage increase	Percentage of full charge*
1τ	63.2%	63.2%
2τ	63.2%	86.5%
3τ	63.2%	95.0%
4τ	63.2%	98.2%
5τ	63.2%	99.3% (taken as full)

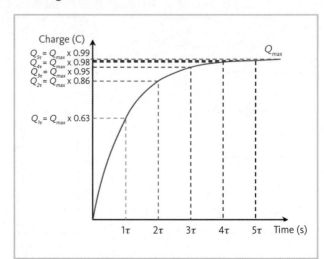

*These values can be calculated using the formula $Q_{n\tau} = Q_{max}(1 - e^{-n})$ but knowledge of it is not expected in Level 3 Physics. However, it is the simplest way to ensure you can accurately plot the points in graph questions, which is required. Sample calculation:

$$Q_{3\tau} = Q_{max}(1 - e^{-3}) = Q_{max} \times 0.950$$

The same exponential relationship applies to the potential difference across the capacitor and the graph is the same shape (see page 297).

As the charge and pd increase, the current decreases, so the graph and values for the current are a reflection of the charge relationship, as shown here.

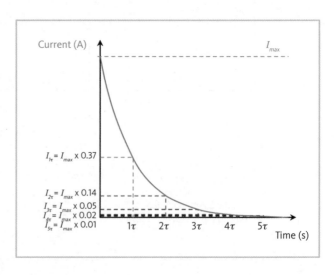

Time	Percentage decrease	Percentage of max current
1τ	63.2%	(100 – 63.2) 36.8%
2τ	63.2%	(100 – 86.5) 13.5%
3τ	63.2%	(100 – 95.0) 5.0%
4τ	63.2%	(100 – 98.2) 1.8%
5τ	63.2%	(100 – 99.3) 0.7% (taken as discharged)

*The values can be calculated using the formula $Q_{n\tau} = Q_{max}(e^{-n})$ but knowledge of it is not expected in Level 3 Physics. Sample calculation:

$$Q_{4\tau} = Q_{max}(e^{-4}) = Q_{max} \times 0.018$$

Worked example: Perfect timing

The graph opposite shows the voltage across a 470 μF capacitor connected in series with a resistor to a battery with an emf of 12.0 V.

a Use the graph to determine the time constant for the circuit.

b Calculate the value of the resistance in the circuit.

c Sketch a line on the graph to show what would happen if the resistance in the circuit doubled.

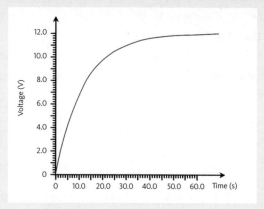

Solution

a

Given	$V_{max} = 12.0$ V
Unknown	$\tau = ?$
Equations	Each time constant represents an increase of 63.2% OR $V_n = V_{max}(1 - e^{-n})$
Substitute	$\dfrac{63.2}{100} \times 12 = 7.584$ V OR $V_{1\tau} = 12(1 - e^{-1}) = 7.585$ V
Solve	Interpolate (red line) across from 7.6 V on the *y*-axis then down to the *x*-axis giving the time constant as $\tau = 12.0$ s

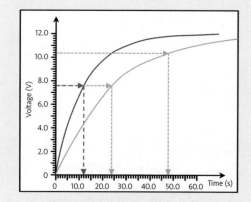

b

Given	$C = 470 \times 10^{-6}$ F, $\tau = 12.0$ s
Unknown	$R = ?$
Equations	$\tau = RC$
Substitute	$R = \dfrac{\tau}{C} = \dfrac{12.0}{470 \times 10^{-6}}$
Solve	$R = 25531.9 = 26$ kΩ (2 sf)

c As $\tau = RC$, doubling R for the same C will result in double the time constant. Sketch in the new points at 63.2% and 86.5% with double the time, then draw in the line (orange).

Discharging

When a charged capacitor is connected across a closed circuit, it will discharge in the opposite way to the charging process. However, if the capacitor is allowed to discharge through a low resistance, the resultant current can be very large.

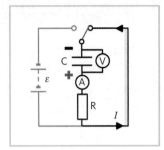

The circuit to the right shows a sample discharge circuit and the graphs below show how the current, voltage and charge change with time.

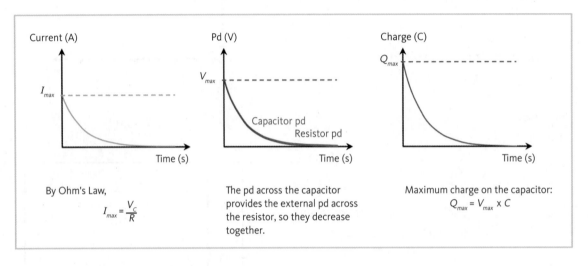

By Ohm's Law, $$I_{max} = \frac{V_C}{R}$$	The pd across the capacitor provides the external pd across the resistor, so they decrease together.	Maximum charge on the capacitor: $$Q_{max} = V_{max} \times C$$

Exercise 6G

1   A security light circuit is designed to turn off after 1.00 minute. This occurs once a 6800 µF capacitor reaches 'full' charge after 5τ.

   a   Calculate the time constant of the circuit. (A)

   b   Determine the size of the resistor required to achieve this turn-off time. (A)

2   A capacitor circuit has a time constant of 171.6 s and is made up of a capacitor connected in series with a 52 kΩ resistor and a 6.0 V battery.

   a   Calculate the capacitance of the capacitor. (A)

   b   Calculate the amount of charge stored on the capacitor when fully charged. (A)

c Determine the maximum current in the circuit, and explain when this will occur. (M)

_____

_____

_____

d Calculate the voltage across the capacitor after the battery has been connected for 343.2 s. (M)

_____

_____

e Explain why the voltage across the capacitor won't have doubled if it is measured 343.2 seconds later (i.e. at 686.4 s). (M)

_____

_____

_____

_____

_____

3 A 220 µF capacitor is connected to a 9.0 V battery until it is fully charged. It is then discharged through a 68.0 kΩ resistor for 44.88 s.

a Determine the maximum charge on the capacitor. (A)

_____

_____

b Calculate the time constant. (A)

_____

_____

c Determine the charge on the capacitor after 44.88 s. (M)

_____

_____

_____

_____

4 The graph shows the voltage across a capacitor, which is discharging through a 28.00 MΩ resistor.

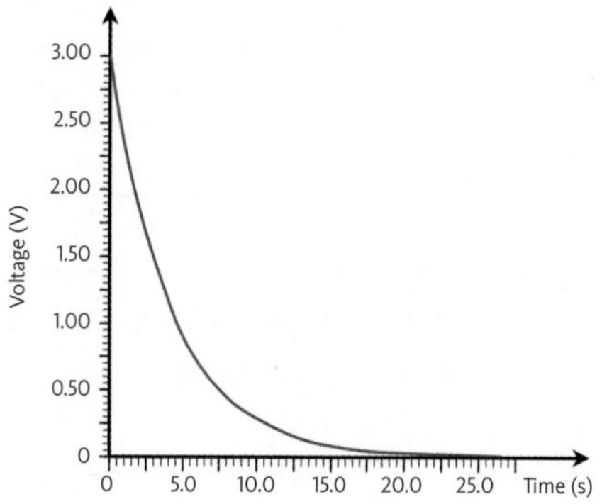

a Calculate the maximum current that flows around the circuit when discharging starts. (A)

_____

_____

b Describe and explain how the current through the resistor will change as the capacitor discharges. (M)

_____

_____

c Use the graph to determine the time constant for the circuit.

_____

_____

d Determine the capacitance of the capacitor in the circuit.

_____

_____

_____

e Calculate the current through the resistor after 5.0 s.

_____

_____

f Determine what would happen to the voltage if the capacitance in the circuit was tripled. Use values to draw a line on the graph to show what would happen.

_____

_____

_____

5 A 150.0 µF capacitor is fully charged by a 3.60 V power supply through a 160 kΩ resistor.

a Determine the maximum charge on the capacitor. (A)

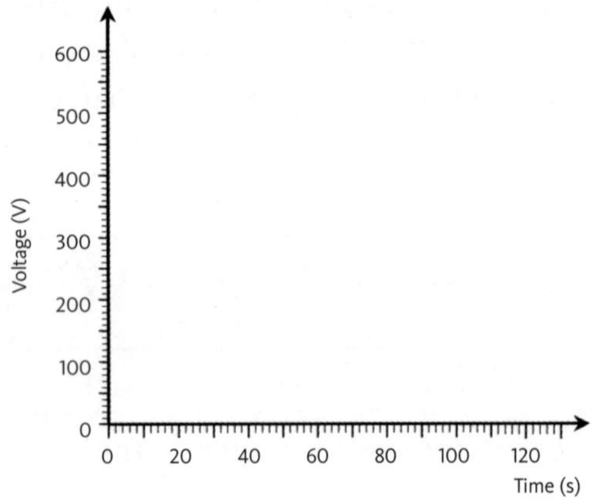

_____

_____

_____

_____

b Determine the time constant for the circuit. (A)

_____

_____

c Plot a graph of charge against time to show the charging progress. (E)

_____

_____

_____

_____

d Use the graph to determine the time taken for the voltage across the capacitor to reach half the emf of the battery. (A)

_____

_____

_____

_____

6.4 Inductance in DC circuits

Faraday's Law and Lenz's Law
Magnetic flux, Φ (phi)

The region around a magnet where magnetic effects can be felt is called a magnetic field. Magnetic flux, Φ, measured in weber (Wb), is a measure of the amount of magnetism in a region. It can be thought of as the number of 'field lines', but remember that field lines do not really exist, and are just an easy way of visualising the amount of magnetism that is present in a region.

Magnetic flux density, B

The strength of a magnetic field at any point is described by the magnetic flux density, B, measured in tesla (T). It is dependent upon the amount of magnetic flux, Φ, passing through a cross-section of area A:

$$B = \frac{\Phi}{A}$$

which states:

$$\text{magnetic flux density (T)} = \frac{\text{magnetic flux (Wb)}}{\text{area (m}^2\text{)}}$$

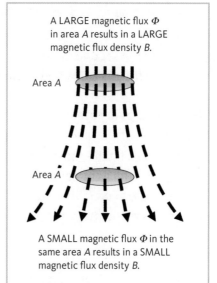

A LARGE magnetic flux Φ in area A results in a LARGE magnetic flux density B.

Area A

Area A

A SMALL magnetic flux Φ in the same area A results in a SMALL magnetic flux density B.

Faraday's Law

When a conductor moves at a steady speed through a magnetic field, it cuts through a magnetic flux, $\Delta\Phi$, in time, Δt, and an emf, ε, is induced in the conductor as a result.

Faraday's law of electromagnetic induction states that:

> The induced emf (V) is proportional to the rate of change of the magnetic flux (Wb s^{-1}).

which states:

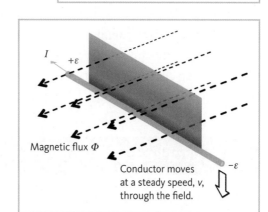

Magnetic flux Φ

Conductor moves at a steady speed, v, through the field.

$$\varepsilon = -\frac{\Delta\Phi}{\Delta t}$$

If the conductor is looped N times through a field then $\varepsilon = -N\dfrac{\Delta\Phi}{\Delta t}$. (See Lenz's law below, about the – sign.)

Lenz's Law

In 1834, the Russian physicist Heinrich Lenz concluded that:

> When an induced current flows, its direction is always such that it will oppose the change in flux that produced it.

Lenz's Law agrees with the **Principle of Conservation of Energy**. Work must be done by some external force against the magnetic force which acts upon the conductor. The energy changed pushing the conductor is transformed into electrical energy and heat in the conductor.

Direction of the induced emf/current

Fleming's right-hand rule

The direction in which the induced emf and **current** around the circuit depends upon:

- the direction of the **magnetic field** provided by the permanent magnet **and**
- the direction the **conductor is moving relative to the magnetic field**.

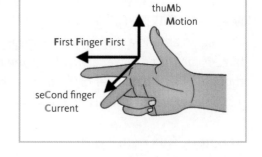

This can be remembered using Fleming's r**I**ght-h**and** rule for **I**nduction.

Remember: the seCond finger refers to the direction of the Current, i.e. positive charge flow. This means that the electrons flow in the opposite direction.

Right-hand grip rule

The direction of the field inside a current carrying coil can be determined using the **right-hand grip rule for solenoids**. If the coil is held in the right hand then:

- the **fingers** curl in the direction of the **current** and
- the **thumb** points in the direction of the **field lines**.

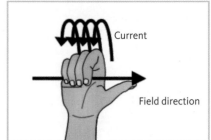

Eddy currents

Whenever a conductor moves in a non-uniform magnetic field, or is placed in a changing magnetic field, emfs are induced in it. These give rise to induced currents, **eddy currents**, which circulate within the body of the metal.

The eddy currents may be large because they follow low-resistance paths, even though the induced emfs are often small. As a result, eddy currents can produce quite considerable heating and magnetic effects and so need to be controlled to avoid wasted energy.

Worked example: An embarrassing relative

Uncle Michael shows a magic trick during Christmas dinner. He drops a small metal cylinder through an aluminium pipe and it falls through very quickly. He then appears to drop the same cylinder through the pipe again but this it takes 10 times longer. Afterwards he states, 'The second cylinder was actually a magnet, but the pipe isn't magnetic, so how does it work?'

With the aid of a diagram, answer Uncle Michael's question.

Magnet moving downwards

Solution

The magnet is moving down BUT it is the **relative motion** of the conductor that is important and must be used in explaining induction effects.

A magnetic field surrounds the falling magnet (this is a 3D field but only the sides have been shown to make it easy to see).

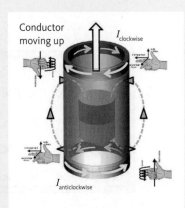

Conductor moving up

$I_{clockwise}$

$I_{anticlockwise}$

At the bottom the fields are pointing outwards, so for a conductor moving upwards Fleming's Right-hand Rule indicates that a current must be flowing anticlockwise at the bottom and clockwise at the top (looking down on the pipe).

Applying the right-hand grip rule to the pipe shows that the current direction causes a magnetic field to develop in the metal beneath the magnet with a North pole pointing upwards repelling the falling magnet.

Above the falling magnet the current direction causes a magnetic field to develop in the metal with a North pole pointing downwards attracting the falling magnet. This is in accordance with Lenz's Law that states that the direction of the induced current is such that it will produce a magnetic field that opposes the motion that induced it.

Exercise 6H

1   In the following situations identify the positive and negative ends of the rod and state the direction, if any, in which the current flows. If no current flows, explain why.

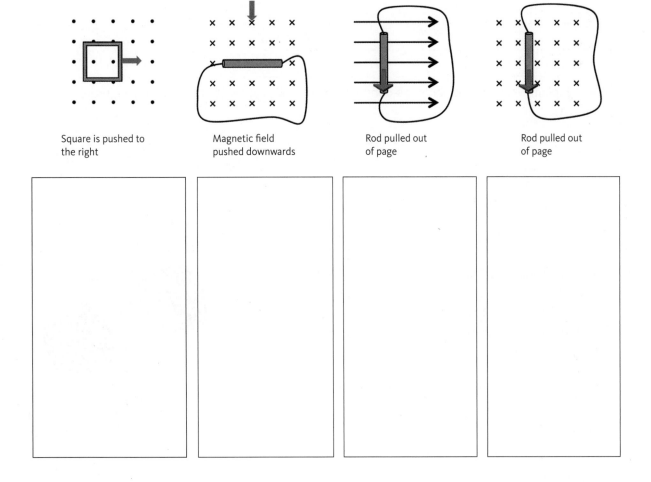

| Square is pushed to the right | Magnetic field pushed downwards | Rod pulled out of page | Rod pulled out of page |

2 As a conductor is pushed horizontally along a metal frame through a magnetic field with a flux of 0.25 Wb, sweeps through the area of 1.44 m² in 3.0 seconds.

    a Calculate the magnetic flux density. (A)

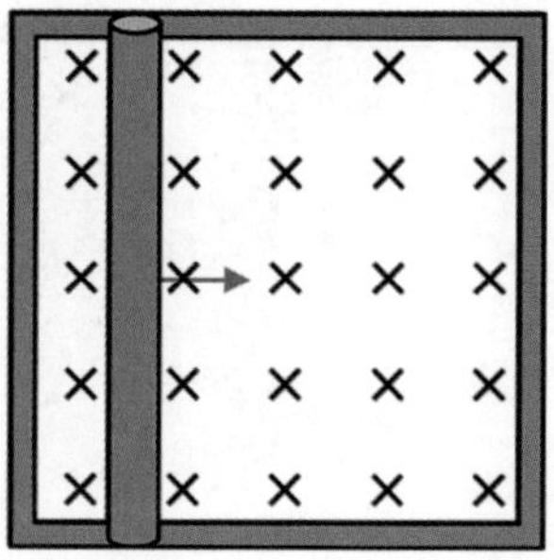

_____

_____

    b Determine the emf induced across the conductor. (A)

_____

_____

    c Indicate on the diagram which end of the conductor becomes positive and explain your reasons. (A)

_____

_____

_____

_____

    d Explain why the conductor must be pushed through the field and can't just be left to roll through. (M)

_____

_____

_____

_____

_____

_____

    e Explain what will happen if the conductor now moves at twice the speed through the field. (A)

_____

_____

_____

_____

_____

3 A strong electromagnet is used to produce a uniform magnetic field of area 1.50×10^{-2} m² and flux density 0.60 T between its poles. A loop of 200 turns of copper wire is placed so that it just surrounds the magnetic field as shown in the diagram. The loop is connected to an external circuit with a 4.8 Ω resistor.

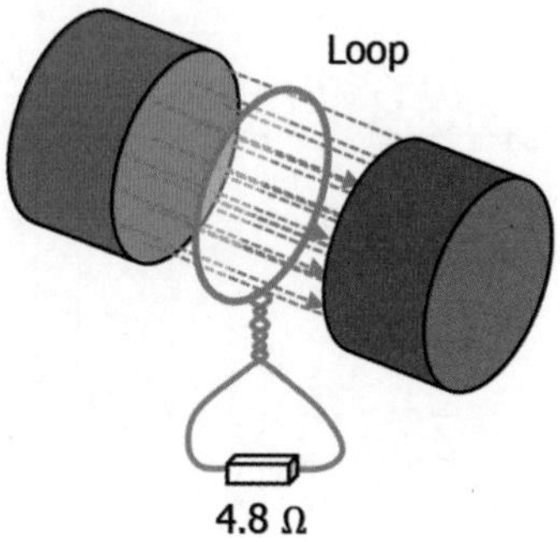

Loop

4.8 Ω

Determine the amount of magnetic flux inside the loop. (M)

_____

a Explain why there is no emf generated inside the loop. (A)

_____

_____

_____

_____

The power supply to the electromagnets is gradually reduced to zero resulting in an emf of 3.6 V being induced in the loop.

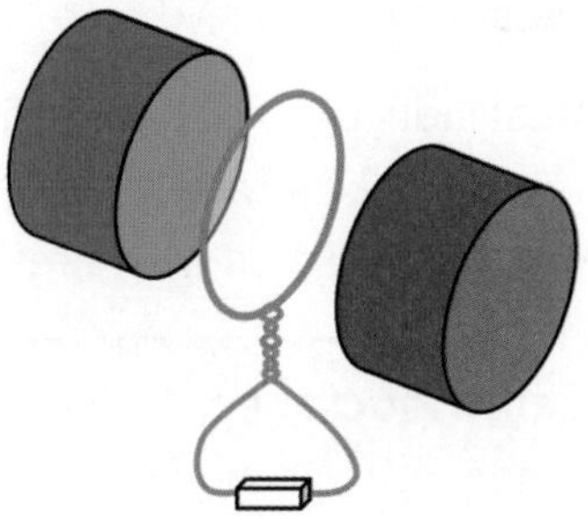

b Calculate the time taken for the power supply to be reduced to zero. (A)

_____

_____

_____

c Calculate the current that flows through the 4.8 Ω resistor. (A)

_____

_____

_____

d On the diagram above, draw the direction of the induced current in the loop and the magnetic field that this current produces. Explain your answer.

_____

_____

_____

_____

_____

_____

_____

_____

4 A freely spinning copper disc is placed in a uniform magnetic field as shown in the diagram. With the aid of the diagram, explain why the disc is rapidly brought to rest even though copper is not magnetic.

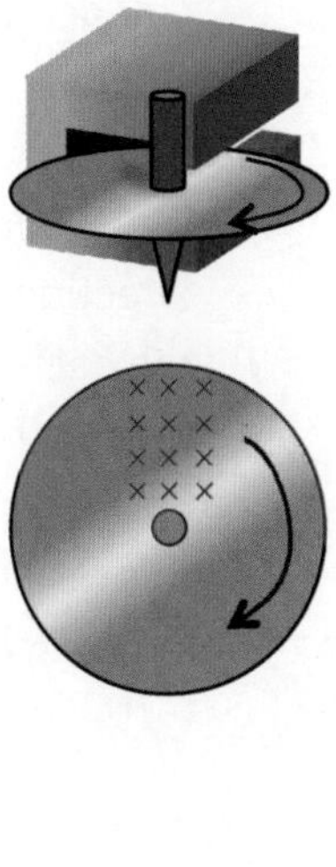

_____

_____

_____

_____

_____

_____

_____

_____

Inductors

Inductance

An inductor is made by wrapping a wire coil around an iron core. When a CHANGING current flows in the wire, it creates a CHANGING magnetic flux. This in turn INDUCES a voltage across the inductor, which opposes the change in the current (Lenz's Law).

Ideal inductors have no resistance to the current flow.

Real inductors will have resistance due to the length of wire used to make the coil. Like a battery that has an internal resistance, the symbol for a **real** inductor is drawn with a resistor in series, enclosed in a dotted box.

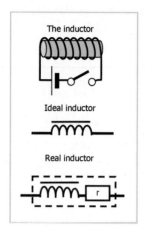

The inductor

Ideal inductor

Real inductor

Self-inductance, *L*

When a circuit containing only **resistors** is switched on or off, the current changes almost instantly. But when a circuit containing an **inductor** is switched on or off, it takes time for the current to change.

This effect is due to the property of **self-inductance**, *L* (measured in henry, H) in the inductor and depends on:

- the number of coils
- the material of the core.

It occurs because a **changing** current, ΔI, causes a **changing** flux, $\Delta \Phi$, which, according to Faraday's Law, induces a **back emf**, ε_b, in the inductor.

$$\varepsilon_b = -L\frac{\Delta I}{\Delta t}$$

By Lenz's Law, this opposes the changing current.

- At switch-on:
  - the back emf opposes the power supply, slowing the increase in current
  - the maximum back emf is limited by the power supply:
    $\varepsilon_b = -\varepsilon_s$ (closed circuit)
  - as the rate of change in current decreases, the back emf decreases until it disappears when the current is stable.
- At switch-off:
  - the emf induced in the inductor now acts in the same direction as the power supply was acting, to oppose the decreasing current
  - the rate of change in current is no longer limited by the power supply so can be very large (open circuit).
  - this results in a large change in flux producing a very large back emf, which can cause sparks as $\varepsilon_b \gg \varepsilon_s$.

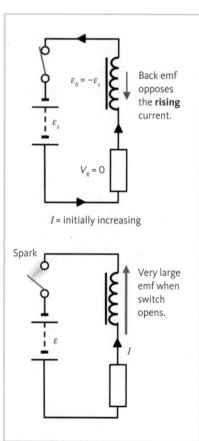

$\varepsilon_b = -\varepsilon_s$

Back emf opposes the **rising** current.

ε_s

$V_R = 0$

I = initially increasing

Spark

Very large emf when switch opens.

ε

I

 ISBN: 9780170368179

Energy stored in an inductor, *E*

When a current, I, flows in an inductor, energy, E, is stored in the magnetic field. The amount of energy stored is dependent on:

- the size of the current, (A)
- the self-inductance, L, of the inductor, (H).

Hence:

$$E = \frac{1}{2}LI^2$$

Time constant, τ

The time it takes for the current through an inductor to reach its maximum value depends on:

- the self-inductance, L, of the inductor and
- the resistance, R, in the circuit,

and is described using the **time constant**, τ, where:

$$\text{time constant (s)} = \frac{\text{self-inductance (H)}}{\text{resistance (}\Omega\text{)}}$$

Expressed mathematically:

$$\tau = \frac{L}{R}$$

The time constant represents the time taken for the current through the inductor to increase to 63.2%, resulting in an exponential relationship that means that the current through an inductor can never reach 100%. The voltage across the inductor decreases exponentially as the current increases.

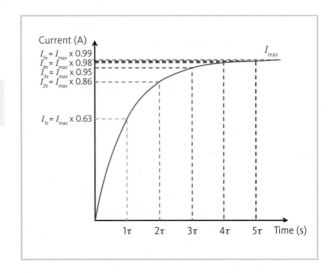

Worked example: Flash of inspiration

A 0.56 H inductor and 220 Ω resistor are connected in series with a switch and a 12 V battery. A spark gap is placed in parallel with the inductor and requires more than 90 V for electricity to arc cross the gap.

The switch is closed and the current rises to a maximum, then the switch is opened and the spark gap flashes. Data loggers are used to measure the current and voltage throughout the experiment and the graphs are shown below.

Explain why the voltage across the inductor is so much greater when the switch is opened compared to when it is closed.

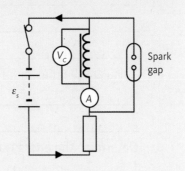

Solution

When the switch is closed, the inductor is in a closed loop with the battery and the resistor, and so the voltages around the circuit must sum to zero (Kirchhoff's Law). This means the maximum voltage across the inductor is limited to the battery pd, hence $\varepsilon_b = -\varepsilon_s$.

The back emf reduces the rate of increase of the current so the back emf gradually decreases until the current reaches its maximum value limited only by the power supply and the resistor; $I = \varepsilon_s/R$.

When the switch is opened, the current falls to zero almost instantly. As $\varepsilon = -L\dfrac{\Delta I}{\Delta t}$, the back emf is very large and acts in the direction of the original current to oppose the decrease. This large emf is sufficient to jump the spark gap and cause a flash.

Exercise 6I

1   A 5.6 H inductor and 22 Ω resistor are connected in series with a switch and a 12 V battery. A neon (non-filament) lamp is placed in parallel with the inductor and resistor, and requires more than 80 V to light. The switch is closed and the current rises to a maximum.

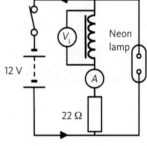

   a   State the back emf across the inductor when it is switched on. (A)

   _____

   b   Calculate the time constant of the circuit. (M)

   _____

   _____

   _____

   c   Calculate the maximum current before the switch is opened. (A)

   _____

   _____

   d   Determine the energy stored on the inductor. (A)

   _____

   _____

The switch is now opened and the current falls to zero in 0.030 s.

e Determine the emf across the inductor and state the direction of the induced current flow through the inductor. (M)

_____

_____

_____

2 A 4.20 H ideal inductor is connected in series to a 6.0 V power supply and a 20.0 Ω resistor.

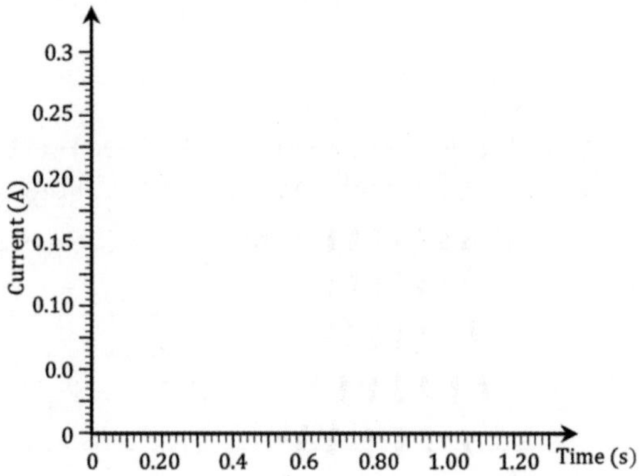

a Determine the maximum current in the circuit. (A)

_____

_____

b Determine the time constant for the circuit. (A)

_____

_____

c Plot a graph of current against time include all essential values. (E)

_____

_____

_____

_____

_____

d Use the graph to determine the time taken for the current to reach half the maximum current. (A)

_____

_____

e Discuss how the behaviour of the circuit would change if you doubled the resistance in the circuit. Sketch the shape of the new I versus t graph above. (E)

_____

_____

_____

_____

_____

3 A real inductor is constructed from 42.0 m of wire coiled into a loop. The wire has a resistance due to its length and is placed in series with a 2.40 V battery and a 2.50 Ω resistor producing a maximum current of 0.75 A. The graph shows how the current through the wire changes with time.

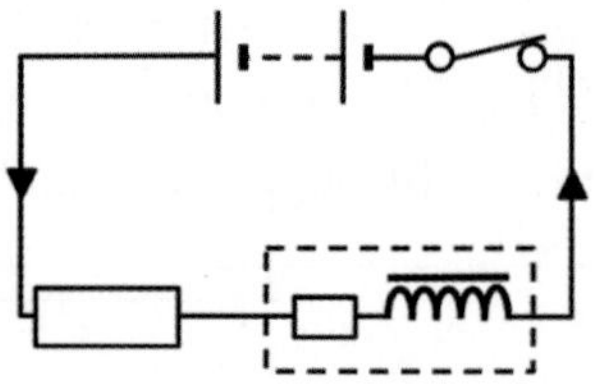

a Use the graph to determine the time constant for the circuit. (A)

_____

_____

_____

b Use the graph to determine the resistance of the wire in the inductor. (A)

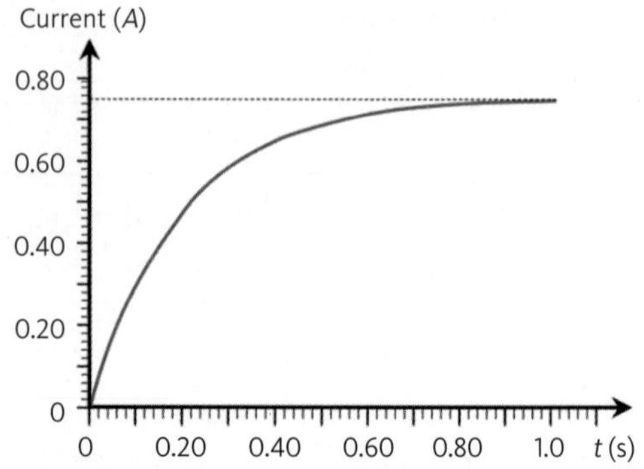

Current (A)

_____

_____

_____

_____

_____

c Show that the inductance of the inductor is 0.64 H. (A)

_____

d Describe and explain how the voltage across the inductor will change as the current rises to its maximum value. Provide supporting values. (M)

_____

_____

_____

_____

_____

_____

e Determine the energy stored on the inductor after 0.30 s. (A)

_____

_____

f A soft iron bar is placed inside the coil of wire. Explain what effect this will have on the time constant and the energy stored in the inductor. No calculations are required.

_____

_____

_____

Transformers

In New Zealand, the mains electricity is delivered around the country as an alternating current (AC) with a voltage of 220 000 V. To make this available for use in homes requires the voltage to be reduced to 240 V. Once in our homes, the voltage must be reduced further for devices like mobile phone chargers and computers.

This can be done using transformers made from two coils of wire wrapped around a common iron core.

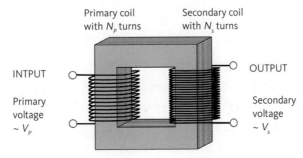

When an AC supply is connected to the primary coil, the changing current in the primary circuit creates a changing magnetic flux in the iron core. This changing flux passes through the secondary coil, where it induces an alternating voltage in the secondary coil. The size of the output voltage and current depends on:

- the input voltage across the primary, V_P
- the number of turns on the primary coil, N_P compared to the secondary coil, N_S.

In an IDEAL transformer the relationship between the voltage and the number of turns is given by:

$$\frac{\text{number of turns on the primary}}{\text{number of turns on the secondary}} = \frac{\text{potential difference across the primary}}{\text{potential difference across the secondary}}$$

Expressed mathematically:

$$\frac{N_P}{N_S} = \frac{V_P}{V_S} \quad \text{and} \quad \frac{N_P}{N_S} = \frac{I_S}{I_P}$$

In an IDEAL transformer no energy is lost (100% efficient) but in REAL transformers some energy is dissipated, mainly in the form of heat.

$$\text{Efficiency} = \frac{\text{power output}}{\text{power input}} \times 100\%$$

Exercise 6J

1 A step-down transformer is used to convert the 240 V AC mains supply for use in a laptop computer. The primary coil has 800 turns and the secondary coil has 60 turns.

   a Calculate the output voltage from the secondary coil. (A)

   _____

   _____

b The input current is 0.375 A. Determine the output current through the secondary. (A)

_____

_____

c A real transformer gets hot in use. Suggest possible causes of this heating effect. (M)

_____

_____

_____

_____

2 Mains power in New Zealand and the US is not the same, so people moving to live in the US often have an issue with using mains-powered devices. To overcome the problem, they can buy a transformer that plugs into the US 110 V AC supply and transforms it to New Zealand 240 V AC, which is required by New Zealand appliances. The primary coil in the transformer has 500 turns on it.

a Calculate the number of turns on the secondary coil to enable the transformer to provide a 240 V output.

_____

_____

b Explain what will happen to the current as a result.

_____

_____

_____

_____

3 In the Manapouri hydroelectric power station, step-up transformers are used to increase the generated voltage from 13.8 kV to a transmission voltage of 220 kV, as it causes the current to decrease by the same factor reducing power loss in the cables.

a Calculate the number of turns on the primary coil if the secondary coil has 1500 turns.

_____

_____

_____

The maximum output power from Manapouri is 600 MW but the transformers are only 75% efficient.

b Calculate the input power to the transformers.

_____

_____

_____

 ISBN: 9780170368179

6.5 AC circuits

AC circuit theory

Electricity is generated by rotating conductors in magnetic fields. As a consequence of the conductor changing direction in the magnetic field, the induced current changes direction. And as the conductor is rotating, the circular motion results in a sine wave output.

AC circuits are described in terms of certain fundamental properties:

- **frequency, f(Hz)**
  The number of oscillations (imaginary revolutions) each second. It is the reciprocal of the time period.
- **angular frequency, ω (s^{-1})**
  This effectively relates the angular velocity of the spinning generator to the linear motion of the current in the wire.

  $$\omega = 2\pi f$$

- **maximum current, I_{max} (A)**
  The maximum current during any cycle. It is represented by the peak of the sine wave.
- **maximum voltage, V_{max} (V)**
  The maximum voltage during any cycle. It is represented by the peak of the sine wave.

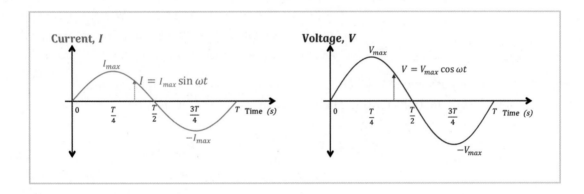

AC and resistors

Power dissipated in resistors

When an alternating current, I, flows through a resistance, R, the power, P, dissipated in the resistor varies through the cycle.

At any instant:

$$P = I^2R \qquad \text{But } I = I_{max}\sin \omega t$$

so

$$P = I_{max}{}^2R\sin^2 \omega t$$

P is always positive because all values of **$\sin^2 \omega t$** lie between 0 and 1 as shown.

So the maximum power is given by:

$$P_{max} = I_{max}{}^2R$$

Average power = ½ max power

$$P_{ave} = \tfrac{1}{2}P_{max}$$
$$P_{ave} = \tfrac{1}{2}I_{max}{}^2R$$

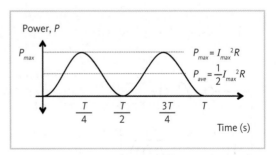

RMS voltage and current

As the AC power is constantly varying, it does not give out as much power as a DC circuit with the same peak values. In the DC circuit below, the DC supply has been set to a steady current, I_{DC}, such that the power dissipated in the resistor is exactly the same as the power dissipated in the AC circuit with the same resistance and voltage.

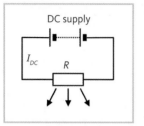

so

$$P_{DC} = P_{AC}$$

$$I_{DC}{}^2 R = \frac{1}{2} I_{max}{}^2 R$$

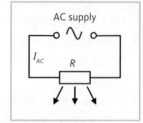

which reveals that the DC current that has the same power output as an AC circuit is given by:

$$I_{DC} = \frac{I_{max}}{\sqrt{2}}$$

I_{DC} is referred to as the root mean square (rms) current, I_{rms}, and allows for direct comparisons between AC and DC circuit. The same is true for voltage.

$$I_{max} = I_{rms} \sqrt{2} \qquad V_{max} = V_{rms} \sqrt{2}$$

Unless stated, you may assume that AC circuit values are rms rather than max values. Ohm's Law and power equations for rms values are identical to the DC versions.

Ohm's Law $\qquad V_{rms} = I_{rms} R \qquad$ and power $\qquad P_{rms} = V_{rms} I_{rms}$

Exercise 6K

1   A kettle is plugged into the AC mains electricity. It is rated as 240 V, 13 A rms.

    a   Determine the maximum voltage and current supplied to the kettle when in operation. (A)

    _____

    _____

    b   Using Ohm's Law, calculate the resistance of the kettle: (A)

        i   using the rms values. _____

        ii   using the max values. _____

    c   Calculate the power output: (A)

        i   using the rms values. _____

        ii   using the average power output.

    _____

 ISBN: 9780170368179

2 A ceiling lamp with a maximum power output of 120.0 W is connected to an AC power supply that produces a maximum current of 7.07 A.

 a Calculate the resistance of the lamp. (A)

_____

 b Calculate the average power output. (A)

_____

 c Calculate the maximum voltage. (A)

_____

 d Determine the rms voltage, current and power ratings for the lamp. (A)

_____

_____

_____

_____

_____

AC and capacitors

When a capacitor is placed in a DC circuit it will gradually become fully charged and prevent any further charge flow in the circuit. When a capacitor is placed in an AC circuit it will start to become charged and oppose charges flowing onto the plates, but when the supply current changes direction, the capacitor is now able to repel charges off in the same direction as the supply current. As a consequence, charges are able to flow in an AC circuit (however no charges ever cross the capacitor).

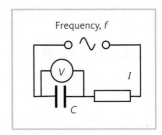

Power, *P*

The energy supplied to a capacitor is stored in the electric field between the plates and then returned to the circuit as the capacitor discharges. This means that **no energy is dissipated** by the capacitor.

Capacitor reactance (X_c)

As capacitors do not dissipate any power, they cannot be described as having a resistance, but capacitors do oppose the flow of charges in an AC circuit. This is described as capacitor reactance, X_c, and is dependent upon the size of the capacitor and the supply frequency.

$$X_C = \frac{1}{2\pi fC} \quad \text{and as } \omega = 2\pi f \text{ so} \quad X_C = \frac{1}{\omega C}$$

From this it can be seen that:

- **low-frequency AC – large reactance**
  The capacitor has time to become charged so will start to oppose the flow of charges in the circuit.
- **high-frequency AC – low reactance**
  The capacitor doesn't have time to store a large charge so offers very little opposition to the flow of charges.

Ohm's Law and reactance

The reactance can also be calculated in terms of the voltage across the capacitor and the current in the circuit using Ohm's Law.

$$V_c = IX_c$$

Impedance in resistor-capacitor (RC) circuits

Impedance, Z, describes the overall effect of the resistance and the reactance when a circuit contains both a resistor and a capacitor.

The resistance is always in phase with the current and the capacitor's reactance lags by 90° ($\pi/2$) so the two quantities must be added using a phasor diagram, Pythagoras's theorem and trigonometry. Remember: The C (sea) is below you. The total impedance/supply voltage is given by: $Z = \sqrt{X_c^2 + R^2}$ and $V_S = \sqrt{V_C^2 + V_R^2}$

The phase angle of the impedance/supply voltage is given by: $\tan\theta = \dfrac{X_c}{R} = \dfrac{V_c}{V_R}$

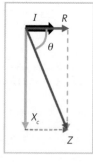

Ohm's Law and impedance

The impedance represents the total resistive effects across the supply voltage so can also be found using Ohm's Law:

$$V_S = IZ$$

Exercise 6L

1   A 680 μF capacitor is connected to a 12.0 V 45 Hz power supply.
    a   Calculate the reactance of the capacitor. (A)

    _____

    b   Determine the rms current in the circuit. (A)

    _____

2   A capacitor is connected to a 24.0 V AC 80 Hz signal generator resulting in a current of 1.33 A.
    a   Calculate the reactance of the capacitor. (A)

    _____

    b   Determine the capacitance of the capacitor. (A)

    _____

    c   Explain how the current in the circuit could be halved. (M)

    _____

    _____

    _____

    _____

3   A 47.0 μF capacitor is connected in series with a 28.0 Ω resistor and a 30.0 V AC 50.0 Hz power supply as shown in the diagram.
    a   Calculate the reactance of the capacitor. (A)

    _____

    _____

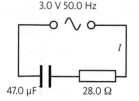

3.0 V 50.0 Hz

I

47.0 μF   28.0 Ω

b Draw a phasor diagram in the space provided and use it to calculate the impedance in the circuit. (E)

_____

_____

_____

c Determine the angle by which the supply voltage lags the current.

_____

_____

_____

d Show that the rms current is 0.409 A. (A)

_____

_____

e Calculate the potential difference across the capacitor and resistor and compare it to the supply voltage. Explain any apparent disagreement with Kirchhoff's voltage law. (E)

_____

_____

_____

_____

_____

_____

_____

4 A parallel circuit is connected up with a large capacitor, resistor, two identical lamps with a similar operating resistance to the resistor, a switch and a 12.0 V DC battery as shown in the circuit diagram.

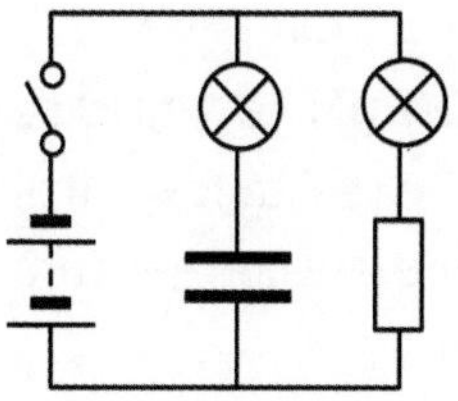

a Describe and explain what happens to the lamps when the switch is closed. (E)

_____

_____

_____

_____

_____

_____

b Describe and explain what would be observed if the battery was now replaced with a high frequency 12.0 V_{rms} AC power supply. (E)

_____

_____

_____

_____

AC and inductors

When an inductor is placed in a DC circuit and switched on, it initially opposes the supply voltage, but as the rate of change of current decreases, the back emf of the inductor decreases until a steady current is reached and the inductor has no effect on the circuit.

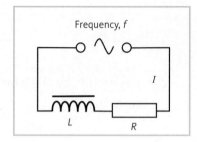

When an inductor is placed in an AC circuit the constantly changing current results in a continuously changing magnetic flux, which induces a back emf that opposes changes in the current.

Power

The energy supplied to an inductor is stored in the magnetic field in the inductor and then returned to the circuit as the magnetic field collapses. This means that **no energy is dissipated** by the inductor.

Inductor reactance (X_L)

As inductors do not dissipate any power, they cannot be described as having a resistance, but inductors do oppose the flow of charges in an AC circuit. This is described as inductor reactance, X_L, and is dependent upon the self-inductance of the inductor and the supply frequency.

$$X_L = 2\pi f L$$

and as $\omega = 2\pi f$ so

$$X_L = \omega L$$

From this it can be seen that:

- **low-frequency AC – low reactance**
  The slow change in frequency results in a small change in flux so a small back emf.
- **high-frequency AC – large reactance**
  The rapid change in frequency results in a large change in flux so a large back emf.

Ohm's Law and reactance

The reactance can also be calculated in terms of the voltage across the inductor and the current through it using Ohm's Law.

$$V_L = IX_L$$

Resistor-inductor (RL) circuits

Impedance, Z, describes the overall effect of the resistance and the reactance when a circuit contains both a resistor and an inductor. As the resistance is always in phase with the current and the inductor reactance leads by 90° ($\pi/2$), the two quantities must be added using vector addition. Remember: The phase diagram looks like an L.

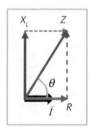

The total impedance/supply voltage is given by: $Z = \sqrt{X_L^2 + R^2}$ and $V_S = \sqrt{V_L^2 + V_R^2}$

The phase angle of the impedance is given by: $\sin\theta = \dfrac{X_L}{R} = \dfrac{V_L}{V_R}$

Ohm's Law and impedance

The impedance represents the total resistive effects across the supply voltage so are related using Ohm's Law:

$$V_S = IZ$$

 ISBN: 9780170368179

Exercise 6M

1 A 0.220 H inductor is connected to a 18.0 V 100 Hz power supply.
   a Calculate the reactance of the inductor. (A)

   _____

   b Determine the rms current in the circuit. (A)

   _____

2 An inductor is connected to a 6.0 V AC 44 Hz signal generator resulting in a current of 0.54 A.
   a Calculate the reactance of the inductor. (A)

   _____

   b Explain what effect tripling the frequency will have on the current. (M)

   _____

   _____

   _____

   _____

   c Determine the inductance of the inductor. (A)

   _____

   _____

3 A 1.30 mH inductor is connected in series with a 3.30 Ω resistor and a 6.0 x 10² Hz power supply providing a current of 2.03 A.

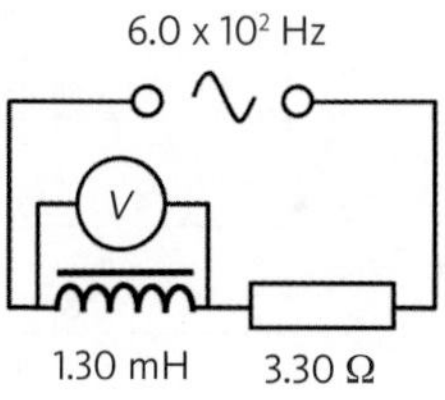

   6.0 x 10² Hz

   1.30 mH   3.30 Ω

   a Calculate the reactance of the inductor. (A)

   _____

   _____

   b Draw a phasor diagram in the space provided and use it to calculate the impedance in the circuit. (M)

   _____

   _____

   _____

   c Determine the supply voltage and the angle by which it leads the current. (M)

   _____

   _____

   _____

   _____

   _____

4   A parallel circuit is connected up with a large inductor, resistor, two identical lamps with a similar operating resistance to the resistor, a switch and a 12.0 V DC battery as shown in the circuit diagram.

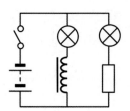

   a   Describe and explain what happens to the lamps when the switch is closed. (E)

_____

_____

_____

_____

_____

_____

_____

_____

_____

   b   Describe and explain what would be observed if the battery was now replaced with a high frequency 12.0 V_{rms} AC power supply. (E)

_____

_____

_____

_____

_____

_____

_____

_____

RCL circuits

When resistors, capacitors and inductors are all included in a circuit, the total reactance must be calculated first. As they are completely out of phase, the total reactance can be determined by finding the difference as follows:

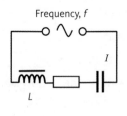

$$X_T = X_L - X_C$$

If the answer is positive, then the circuit is described as being **inductive**.
If the answer is negative, then the circuit us described as being **capacitive**.

The total reactance can then be added to the resistance using a phasor diagram, Pythagoras's theorem and trigonometry.

$$Z = \sqrt{X_T^2 + R^2}$$

   ISBN: 9780170368179

Resonance

As the supply frequency, f, in an LCR circuit increases, X_L increases but X_C decreases.

At a certain frequency, called the resonant frequency, f_r: $X_L = X_C$

as $X_L = 2\pi f_r L$ and $X_C = \dfrac{1}{2\pi f_r C}$ so $2\pi f L = \dfrac{1}{2\pi f C}$, which can be rearranged to give:

$$f_r = \frac{1}{2\pi\sqrt{LC}}$$

At the resonant frequency, f_r:

- the two reactances cancel out due to being in opposite phase, so the impedance is at a minimum as it is only due to the resistance, and
- the current is at the maximum for the circuit.

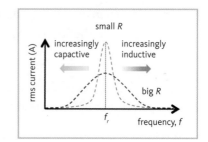

GUESSing – LCR circuit solver

LCR circuits can seem overwhelming with all the different quantities and phasors so it is important to apply the GUESS strategy and use two 'K' diagrams to help organise the quantities.

Step 1:

- As soon as you know you are dealing with an AC circuit question, draw two 'K' diagrams (colour is optional).
- If you know the circuit is inductive, draw the purple arrow up, otherwise just draw it as shown, and correct it later.
- The current arrow is a different style to make it stand out.

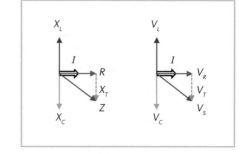

Step 2:

- Label one with resistance, reactance and impedance.
- Label the other with voltages.

Step 3:

- Write in all the values that are **G**iven, identify the **U**nknown and use **E**quations to **S**ubstitute in any values you can quickly determine, e.g. X_T, they may come in handy later.
- Now you're ready to start **S**olving the question.

Exercise 6N

1   An 80.0 Ω resistor is connected in series to a 24 µF capacitor, an inductor with a reactance of 31.4 Ω and a 240 V AC 50.0 Hz power supply.

240 V AC 50 Hz

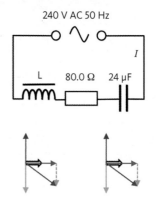

 a   Show that the reactance of the capacitor is 132.6 Ω. (A)

_____

_____

 b   Calculate the inductance of the inductor. (A)

_____

_____

c Determine the total reactance of the circuit and state whether it is capacitive or inductive. (M)

_____

_____

_____

d Draw a phasor diagram in the space provided and use it to calculate the total impedance in the circuit. (M)

_____

_____

e Determine the current in the circuit. (A)

_____

_____

f Determine the frequency at which the circuit will resonate. (A)

_____

_____

g Calculate the new current when the circuit is at resonance and explain why this is the maximum current possible in the circuit. (M)

_____

_____

_____

_____

2 A 127 Ω resistor is connected in series to a capacitor, an inductor with an inductance of 0.20 H and a 30.0 V aC 1.60×10^2 Hz signal generator. The supply voltage leads the current by 41.06°.

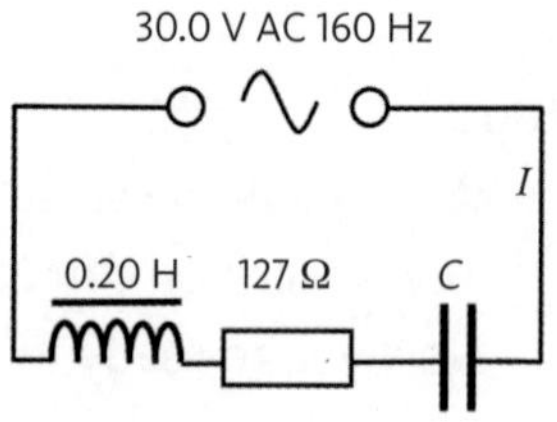

30.0 V AC 160 Hz

0.20 H   127 Ω C

I

a Is the circuit inductive or capacitive? (A)

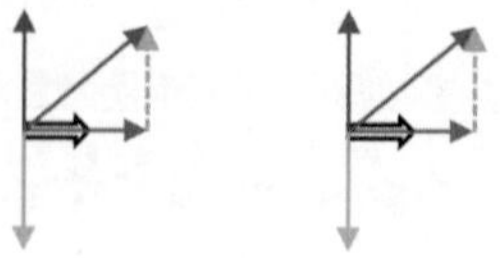

_____

_____

b With the aid of a phasor diagram, calculate the total impedance of the circuit. (M)

_____

_____

_____

_____

c   Determine the total reactance of the circuit and hence find the capacitance of the capacitor. (E)

_____

_____

_____

_____

_____

The frequency of the supply is now altered until the voltage across the inductor and the capacitor are identical at 31.85 V.

d   Determine the frequency of the supply. Clearly explain your solution. (M)

_____

_____

_____

e   Calculate the current in the circuit when at resonance. (A)

_____

_____

f   Explain how swapping the resistor for a 12.7 Ω resistor will affect the behaviour of the circuit when at resonance. (M)

_____

_____

_____

3   A resistor is connected in series with a variable capacitor, an inductor and a 2.00×10^3 Hz signal generator. The voltage across the inductor is 58.91 V, across the resistor is 12.11 V and across the capacitor is 86.36 V.

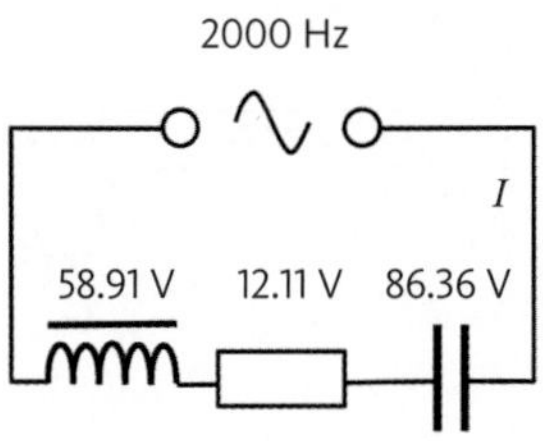

2000 Hz

I

58.91 V   12.11 V   86.36 V

a   Is the circuit inductive or capacitive? (A)

_____

_____

b   With the aid of a phasor diagram, calculate the rms voltage of the power supply and its phase angle. (M)

_____

_____

_____

_____

_____

_____

_____

A current of 0.20 A is measured in the circuit while operating at 2.00×10^3 Hz.

c Calculate the capacitance of the variable capacitor. (M)

_____

_____

_____

d Explain how the capacitance needs to change to bring the circuit to resonance. (E)

_____

_____

_____

_____

_____

_____

_____

_____

e Calculate the reactance of the capacitor when at resonance. (M)

_____

_____

_____

_____

_____

_____

 ISBN: 9780170368179